THE NEW NATURALIST

A SURVEY OF BRITISH NATURAL HISTORY

MAN AND THE LAND

The aim of this series is to interest the general reader in the wild life of Britain by recapturing the enquiring spirit of the old naturalists. The Editors believe that the natural pride of the British public in the native fauna and flora, to which must be added concern for their conservation, is best fostered by maintaining a high standard of accuracy combined with clarity of exposition in presenting the results of modern scientific research. The plants and animals are described in relation to their homes and habits and are portrayed in the full beauty of their natural colours, by the latest methods of colour photography and reproduction.

THE NEW NATURALIST

MAN AND THE LAND

by

L. DUDLEY STAMP

WITH 35 COLOUR PHOTOGRAPHS
46 PHOTOGRAPHS IN BLACK AND WHITE
52 ILLUSTRATIONS IN THE TEXT

COLLINS
14 ST JAMES'S PLACE, LONDON

First published in 1955
Second Edition, revised, 1964
Third Edition 1969

A / 333. 70942

CONTENTS

PLATES IN COLOUR

It should be noted that throughout this book Plate numbers in Arabic figures refer to Colour Plates, while Roman numerals are used for Black-and-White Plates.

PLATES IN BLACK AND WHITE

EDITORS' PREFACE

DUDLEY STAMP's first book in the *New Naturalist* Series, *Britain's Structure and Scenery*, has proved, as his fellow Editors always suspected it would, to have been accepted by the public as a fundamental contribution to the geography of our Islands. Nobody knows exactly what geography is. All we can say is that to take a narrow view of the study of the earth's surface is incorrect, and that the true geographer must not only be a geologist and a geomorphologist, but an historian and a liberal naturalist. Readers of *Britain's Structure and Scenery* must know Dudley Stamp as all these things. Readers of *Man and the Land* will find his many-sidedness abundantly confirmed. If asked his business, Dudley Stamp would probably say that it was geography and that ecology was his recreation. It seems to be a businesslike recreation.

Throughout the *New Naturalist* Series our policy has been to regard man and his works as an integral part of Britain's natural history. The Series could not be complete without the direct assessment of man's influence on the face of the land, on its animal and plant communities, that we offer here.

Since 1930 Dudley Stamp has been Director of the Land Utilisation Survey of Britain. This organisation, which he founded, mapped the land use of every acre in the 1930's, an astonishing task which has formed the basis of Britain's planning for its agricultural future. Since 1942 he has been Chief Adviser on Rural Land Use to the Minister of Agriculture. In 1945 he became a Professor of Geography in the University of London, and since 1948 he has been University Professor of Social Geography at that University (at The London School of Economics). In this Research Chair much of his work has been devoted to the scenery of our land as altered by man, the social animal.

Readers of *Man and the Land* will discover a vast store of personal field experience behind the synthesis. There are few parts of the British Isles that our energetic fellow-Editor has not visited, with his penetrating geographer's eye, his flair for both detail and for generalisation, and, we might add, his camera.

What emerges from this remarkable book is a new and unusual portrait of our islands, and the very clear conclusion that we in Britain are many times fortunate in our land and climate. Britain's land has been able to withstand an exploitation that goes as far back into history as almost any other. It has kept its fertility, and its personality, through years of human increase, so that even with its teeming population of today, which has to buy half of its food abroad, its land remains green, pleasant, and natural. Man has made many terrible mistakes, and may make worse ones, but Professor Stamp convinces us that we certainly have the knowledge, and may even have the wisdom to preserve our wonderful natural resources from decay, impoverishment, and disaster.

THE OTHER EDITORS

AUTHOR'S PREFACE

THIS BOOK may fairly be regarded as a natural sequel to my previous volume in the New Naturalist Series—*Britain's Structure and Scenery*. I was then concerned to demonstrate something of our present day debt to the countless aeons of geological evolution, to show how the form of our land with its rugged mountains, its rounded hills, its smoothed river valleys and waterside flats was the result of the action and interaction of many factors, but especially the nature of the underlying rocks and the process of weathering in our cool but humid climate. My object was to look below the surface : to see the influence of structure on scenery.

In the present book I am taking all that for granted and I am looking at the surface and trying to trace there the hand of man in shaping the scenery as we actually observe it today. The inevitable conclusion is reached that very little indeed of the scenery, even in the wildest and least touched parts of our islands, can truthfully be described as natural.

Over the past two millennia—ever since man began in earnest to clear the natural vegetation to make way for the crops and animals of his own choice—our scenery has undergone a process of gradual evolution. The process is still going on, it will continue to go on. Our motives today are largely economic and perhaps the age of the great landscape architects of the late 18th century was the only period when there was a conscious attempt deliberately to alter scenery for its own sake. At that time the great landowners had no inhibitions about preservation : they were determined to create scenery and, according to the fashion of their time, to create beauty. In these days we are less sure of ourselves : consciously or unconsciously we fear the rising tide of change and so seek to crystallize and preserve for ever, or at least for the foreseeable future, a particular stage in a natural evolution. But our land is clothed with a living mantle and tenanted by living creatures. All living things are born, mature, decay and die, to be replaced in due course by others which are different. However much therefore we may enjoy a particular phase of scenic evolution it is not a conversation piece to be framed and hung upon the walls of a museum or art gallery. It is alive, and our task is so to direct its growth that we create beauty, not destroy it.

There is another aspect of the living if man-made scene. If one dare postulate a "typical man" in Britain, in nine cases out of ten he would be a townsman. For some 48 or 50 weeks out of every 52 he will be seeing nothing but scenery which is essentially artificial : even when he escapes on vacation he may still see little that is truly natural. For every acre of Britain's land which is clothed with vegetation as nature would have it there are a hundred acres whose plant-cover has been determined more or less directly by the hand of man. For every wild mammal he is privileged to see he sees a hundred horses, cows, sheep and pigs. It would indeed be interesting to ascertain what proportion of our population has never seen a native wild mammal other than a bat—for even the familiar rabbits, rats, mice and grey squirrels are aliens and a feature of man-made artificial environments.

No excuse seems therefore to be needed for considering in this book those features of the landscape which make up its greater part for the average man—the crop plants of the field, the trees of orchard and plantation, the animals and birds of the farm, as well as those aliens in the plant and animal worlds which have made Britain their home.

I am fully aware of the extreme difficulty and complexity of the task I have undertaken. I have trespassed into many specialist fields and have rushed in where angels might fear to tread. In every chapter experts will find their own specialisms cursorily surveyed : in the sections covered by Chapters 2 to 6 for example I have had perforce, to make my story complete, to tread well-worn though thorny paths of British history. As soon as one deals with the evolution of even one aspect of Britain through the ages, a vast literature is involved and the task of selection is a very difficult one. So, throughout, I have asked myself the question : what in field and farm, in the humanized landscape of this country, does the observer see, the origin of which he would like explained, in so far as explanation is possible ?

So my story is the elimination of wild nature at the hand of man over the ages, and the gradual substitution of a landscape largely of his making, clothed with trees and plants of his choice, and tenanted by animals which he has either tended or tolerated.

Handed down to us from the days when paper was scarce, unknown, or not to be trusted for permanent use are those records inscribed on parchment from which previous writings had been carefully though not always fully erased. Most often the revelations of such palimpsests are in the fragments which remain of the older work : well has it been claimed that the most complex and exciting of all palimpsests is the map of Britain from which can be deciphered, though only by the exercise of much skill and patience, a whole succession of past records of human occupance and land use. One should perhaps say the surface of Britain itself since no map has ever been produced, or indeed could be produced, to show all the superimposed records which the land surface itself enshrines. Some records—the ridges of once ploughed, now grassed, Anglo-Saxon fields for example—stand out for all to see ; others, like the traces of small Celtic fields, cannot be seen from the ground but only from the air.

Naturally many of the relict features are capable of more than one interpretation. Take for example a landscape where scattered trees are more numerous than elsewhere. Is one viewing the ultimate remnants of what was once a "forest" consecrated to the art of hunting and the use of the privileged ; is one seeing the results of some past landowner's concept of beauty and design in landscape ; or are the isolated trees the result of survival in face of combined physical and biotic factors ?

I have been encouraged to venture into these fields of landscape interpretation because there is not yet by any means unanimity of opinion amongst the experts. On the one hand are those who see the pattern of rural England firmly established in its main lines by Anglo-Saxon times with the majority of our villages already established, parish bounds agreed, county and shire already with roughly the limits they have today. Some historians have gone so far as to state that "by the end of the thirteenth century there was more land under the plough than ever before or possibly since" (Myers, 1952)—which would mean more than 18,000,000 acres in England alone. Others would claim that as late as 1485, "We have to picture the countryside as a sort of agrarian archipelago, with innumerable islands of cultivation set in a sea of "waste" (Bindoff, 1950).

In connection with these differing interpretations one striking fact has emerged from my review of the historical development of the cultural landscape. It is at least startling to me because it was unexpected, and I think may prove of some interest to historians. At the present time using all the techniques known to modern agricultural science which result in a level of production per unit area in Lowland Britain second only to a few even more highly developed areas, we are able to produce 55 per cent. of the food consumed—in other words to feed from the output of our own land 55 per cent. of the existing population. In England and Wales we could at present feed 24,000,000 people from the output of our 24,000,000 acres of crops and grass. It is the familiar story that despite successive technological advances food production has not kept pace with population growth. In the not so distant past England was a corn-exporting country as well as a wool-exporter. At some stage or stages therefore, home consumption of agricultural produce overtook home production. Allowing however for the inefficiency of early and mediaeval systems of farming, as well as a lower standard of living, the conclusion seems to be that for centuries, indeed probably for 1,500 years, the population of England has been at or near the maximum which the home agriculture could support—and that throughout the centuries there has been an ever-present pressure of population on land. Malthusian principles, with certain modifications, have been operating virtually since Roman times. Although much land may have been less efficiently used in the past because of a lack of technical knowledge and skill, I find the conclusion inevitable that much misunderstanding has arisen because of the use of the word waste. For more than a thousand years there has been very little of the surface of Britain, especially Lowland Britain, which has not been playing its part in the national economy.

Because the evolution of land use must be a continuous process, I make no excuse for linking in my last three chapters past, present and future. The comprehensive land use planning upon which we, as a nation, have embarked is doomed to fail unless we understand the present position and the factors which have led up to it.

In my incursions into fields of knowledge in which I am a stranger—necessitated to secure a rounded picture of the whole—I have sought the guidance of many long-suffering friends and colleagues whose help is gratefully acknowledged and is recorded against the chapters concerned. In addition I am greatly indebted to those who have read through the typescript as a whole—especially to my colleagues Professor R. O. Buchanan and Professor T. S. Ashton and to my fellow editors James Fisher and John Gilmour. I should like also to thank Mrs. E. Wilson for her skill and patience in preparing the text-illustrations.

NOTE TO THE SECOND EDITION

I have taken this opportunity of meeting various points raised by reviewers, of noting some of the many developments of the past nine years and of bringing many statistics up-to-date.

October, 1963   L.D.S.

CHAPTER I

IN THE BEGINNING — THE BRITISH EDEN

SINCE THIS BOOK is to cover the story of man's interference with nature in Britain it ought strictly to begin with the first appearance on the scene of man himself. If however we were to go to the Pliocene ancestors of man it would take us back the greater part of a million years. The first appearance of primitive man was destined to be followed by the repeated onslaughts of the Great Ice Age when, four times at least, the ice sheets waxed as the cold intensified or waned as the milder inter-glacial periods intervened. Man left his river bank settlements : retreated into caves : emerged again as the ages passed. Beetle-browed Mousterian man disappeared before the superior wisdom and skill of the direct ancestors of our own species, *Homo sapiens*. The Fourth and last glaciation was probably at its height about 70,000 B.C. and by that time *Homo sapiens* had appeared. There seems to have been a last recrudescence of extreme cold about 25,000 B.C. so that the beginning of the final retreat of the ice dates only from some 20,000 years ago. This date is often used by geologists as marking the end of the Pleistocene and the beginning of the Holocene or definitely post-glacial period.

At first, quite naturally, Britain had an ice-margin type of climate—cold and probably windswept with a tundra-like vegetation cover supporting herds of reindeer. Later the summers became warmer and the land may have resembled more closely the steppelands of Siberia. Today we have some survivals of the steppe flora if little of a fauna which was dominated amongst mammals by a wild horse.

Just as today there is a belt of scrub-birch on the colder poleward side of the coniferous forests, so forests of birch spread over the country. There were climatic oscillations but there followed generally speaking a cold if rather dry period favouring the invasion of the British area and its colonization by Scots Pine. The country was at that time joined by land bridges to the continent of Europe and hazel followed the pine.

The detailed sequence of post-glacial times is much more accurately known than it was a few decades ago, thanks to the long-continued and detailed studies of the palynologists, or students of fossil pollen—work initiated by the Swedish geologist Lennart von Post and the naturalist Lagerheim and applied in this country, especially by the Cambridge School of Botanists led by Dr. H. Godwin, since the 1920's. The pollen grains of many plants have characteristic features by which they are readily recognized and are so nearly indestructible as to have been preserved over thousands of years in peat deposits. Though peat is derived mainly from bog and swamp plants, it includes much pollen blown from afar, especially tree pollen which is liberated at a considerable height from the ground and readily distributed by wind. Whereas earlier workers had to rely on the sequence in peat deposits disclosed to the naked eye or the simple lens and could determine that what are now wet bog-covered lands were at several previous periods dry enough to support forest growth, the modern workers can determine layer by layer the proportion of different pollens and so reconstruct the vegetation over a wide area around. It is found that the same sequence is repeated in different localities in the British Isles and over the continent of Europe and each characteristic pollen assemblage can be correlated with definite climatic conditions.

Although the *sequence* is the same, any given pollen-assemblage in Britain is not necessarily contemporaneous with a similar assemblage in a country from which the ice retreated later—for example Sweden. Much confusion has indeed arisen because this simple fact has been overlooked. British deposits can now be dated approximately in years but the geochronology is somewhat different from that established by De Geer in Sweden.

Evidence of human activity in the form of man's tools and weapons of stone and later of metal and occasional though rare remains of man himself permit a correlation to be made with cultural or archaeological periods. The evidence has already been reviewed from different points of view by the authors of previous volumes in the New Naturalist Series. I considered it briefly from the point of view of geological history in *Britain's Structure and Scenery* (Chapter 14) ; Gordon Manley enlarged on the climatic fluctuations in his *Climate and the British Scene* (Chapter 12) ; W. B. Turrill (*British Plant Life,* Chapter 5), W. H. Pearsall (*Mountains and Moorlands,* Chapter 10) and John Gilmour and Max Walters (*Wild Flowers,* Chapter 4) considered it in relation to the origin and evolution of the British flora ; L. Harrison Matthews (*British Mammals,* Chapter 13) in connection with the origin of the British mammalian fauna. All these authors accept the general sequence which has been established and which is expressed succinctly in the following table based on H. Godwin, *The History of the British Flora*, C.U.P., 1956 (p. 62).

TIME-SCALE		CLIMATIC PERIODS	POLLEN ZONES	VEGETATION	ARCHAEOLOGICAL PERIODS	
A.D.	2,000	Recent (warmer and drier)				
	1,000		VIII	Alder-birch-oak with some beech	Romano-British	
	0	Sub-Atlantic (cool and wet)			Iron Age	
						500 B.C.
B.C.	1,000	Sub-Boreal		⎫ Alder-Mixed	Bronze Age	
	2,000	(drier)	VIIb	⎬	Neolithic	
	3,000			⎭ Oak Forest		3,000 B.C.
	4,000	Atlantic (warm and wet)	VIIa	⎫		
	5,000			⎬	Mesolithic	5,000 B.C.
		Boreal	VI	Hazel-pine	⎬	
	6,000	(warm and dry)	V	Hazel-birch-pine		
	7,000					
		Pre-Boreal	IV	Birch	⎭	
	8,000					8,000 B.C.
	9,000	Upper Dryas	III	Tundra-steppe	⎫	
	10,000	Allerød	II	Birch woods	⎬ Upper	
	11,000	Lower	I	Retreat of	Palæolithic	10,000 B.C.
	12,000	Dryas		glaciers	⎭	

(Pollen zones I–III marked *Late-Glacial*)

The above table coincides closely with the details of post-glacial climatic fluctuations described by Manley.

It must not indeed be thought that we have yet anything like the whole story. More is now known of the sequence of arboreal vegetation than of the lowly vegetation cover of flowering plants, grasses, ferns and mosses. The relative proportion of non-tree pollen however, indicates openness or otherwise of the vegetation cover. Though plant sequences can be related to climatic pulsations we have still to fit in the geologists' and the geomorphologists' pieces of the puzzle. In Scandinavia, as the great weight of the huge ice sheet was removed, the land actually rose. Did the Highlands of Scotland do the same when the British ice-caps melted away? When the ice sheets melted vast bodies of water were added to the oceans and the water level rose so that low-lying sections of neighbouring lands were flooded. Conversely when great ice sheets were re-formed water must have been withdrawn from the oceans so that shallow seas became land. Certainly within the last 15,000 years forests have flourished on lands now covered by waters of the North Sea : land bridges have permitted the migration of both plants and animals where now seas forbid their passage.

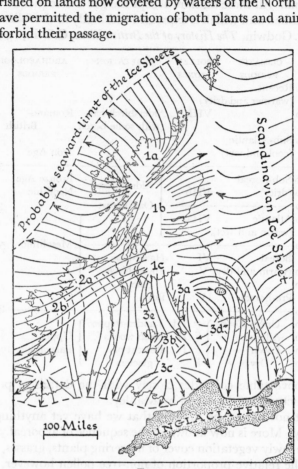

FIG. 1

Map of the British Isles showing the maximum extension of the ice sheets. A pre-existing flora and fauna could only have survived under the tundra conditions of the unglaciated sections

So far as Britain is concerned the last glaciation is the important one because it almost wiped the slate clean. It is highly probable that only the lowly plants of a tundra vegetation survived over the unglaciated lands of southern England and very few, if any, of the larger animals. It is only therefore with late glacial and post-glacial times that we need be concerned.

LATE GLACIAL TIMES (before 12,000 B.C.)

As the ice sheet retreated from northern Britain with halts and brief re-advances so a tundra vegetation became established over the country, with arctic birches slowly advancing from the continent into the south-east. The climate is described by Manley as having been cloudy, windy and raw, interrupted by the milder Allerød period. This was the period when some of the widespread plants now found at highest levels established themselves in the country.

When the ice disappeared from the lowlands it left behind an irregular mantle of glacial drift—boulder clay of varying composition, spreads of gravels and sands with morainic ridges to mark halts in the retreat of the ice and outwash fans splaying from them. The development of the ice sheets had interfered with pre-existing drainage: the surface left was irregularly and often ill drained. There were stretches of shallow water: glacial lakes in which were deposited sheets of sand, silt and mud. From some of these lakes the water in due course found an exit and drained away to leave potentially fertile plains which today play an important role in the country's agricultural economy. Others were too low-lying for the water to be able to escape, or the water occupied depressions with an impervious floor of boulder clay. They became the home of water-loving plants and such is the origin of some of our bogs and mosses such as those of the Lancastrian plain; others lie on ill-drained marine clays. Some meres are the last remnants of glacial lakes.

So the first picture is one of a tundra landscape with innumerable lakes and bogs over the lowlands: uplands swept clear of soil and displaying bare ribs of rock between the lichen-clad slopes. Though the south of England—south of a line joining the Thames and the Bristol Channel—was never glaciated in the sense of being covered by ice sheets, the surfaces must have been covered by frozen rock-sludge. When the ice retreated and thaws set in this sludge slid downwards and resulted in such interesting deposits as the so-called coombe rock which clogs many of the valleys in the chalklands. The chalklands themselves, over large stretches their soft white limestone naked to the elements, probably afforded some of the best drained land of Lowland Britain, though much wetter than at the present day.

LATE GLACIAL TIMES (12,000 to 8,000 B.C.) and THE PRE-BOREAL PERIOD

The amelioration of the climate was not progressive. A period when spring and early summer were relatively dry and when the July temperature over what are now the Midlands averaged 45°F. according to Manley, gave place to the 'Allerød phase' of distinctly warmer conditions when birches of tree size reached Berwickshire and the Lake District and there were probably no glaciers remaining south of Scotland. But then followed a deterioration: the birches retreated southwards, tundra was re-established in the Lake District and North Wales and there was a general re-advance of ice in Scotland. Manley gives the dates 8,300 to 7,800 B.C. for this period of deterioration. It is shown in Godwin's table (p. 3 above) as Upper Dryas extending from 8,800 to 8,000 B.C.

During these times the level of the sea relative to the land was much lower than at the present day. A great plain occupied what is now the floor of the southern half of the North Sea and over it meandered the Rhine and its tributaries, one of which was the ancestor of the Thames. Forests flourished over part at least of this plain. This has long been known from the fragments of peat or "moorlog" not infrequently brought up from the floor of the North Sea in trawl nets, but it was left to pollen analysis to show that Zones IV, V and VI were all represented in the fragments. It does mean, of course, that during these millennia Britain was just a part of the landmass of continental Europe, that plants and animals were able freely to migrate into the British area in the wake of the disappearing ice. Apart from the well-established birch and pine and willows on wet ground around the meres, in some places small patches of hazel, oak, elm and alder had already appeared.

THE BOREAL PERIOD

The climate had been growing steadily warmer and drier and in the two thousand years of the Boreal Period, 7,000 B.C. to 5,000 B.C. it was actually warmer than at the present day as evidenced by the greater northern extension of a number of warmth-loving plants. Pine forest occupied much of England and stretched also into Ireland and Scotland where pines vied with but were never as abundant as birch. The existence of this birch zone north of the coniferous zone is of course paralleled in northern Europe at the present day. Hazel (*Corylus*) had also invaded Britain in force and formed great thickets so that the Boreal Period is Godwin's Pollen Zone V of Hazel-Birch-Pine passing into Zone VI of Pine-Hazel. The country at this time was inhabited by Mesolithic peoples who, like their Palæolithic predecessors, were hunters and not agriculturalists and so had little effect on the natural vegetation.

During the whole of the Boreal period the land was steadily sinking

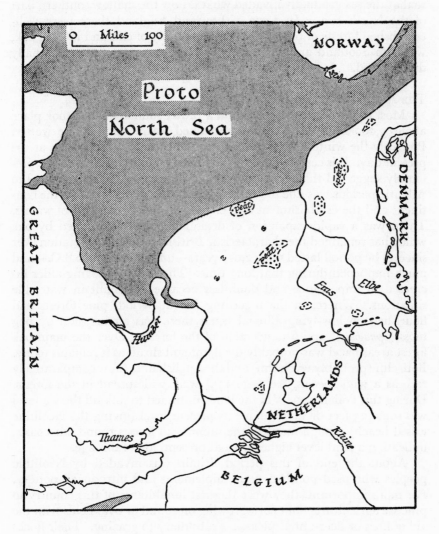

FIG. 2
The Geography of Pre-Boreal Times, about 7,500 B.C.

so that the sea gradually invaded what is now the shallow southern part of the North Sea. In parts of Lowland England there were well-established forests of broadleaved trees notably of elm and later of oak, whilst in Scotland and the north the last remnants of glaciers finally disappeared.

THE ATLANTIC PERIOD

Most authorities are agreed that a rather sudden change took place about 5,000 B.C. The climate remained warm but became wetter. Probably the winters were milder and the climate moister than at the present day. It is certain that the land had been sinking and I have previously suggested that the suddenness of the change from the Boreal to Atlantic periods was due to the final separation of Britain from the continent and the establishment of a circum-insular circulation of waters. There was a rapid expansion of deciduous forest dominated by oak which has remained the characteristic British woodland vegetation ever since. The period lasted some 2,000 years—until about 3,000 B.C.—and much happened during that long time. The moisture-loving alder increased in importance and doubtless became dominant in waterside situations. Where, in the preceding Boreal period, pine forests had flourished on low-lying alluvial tracts there was an invasion by bog moss (*Sphagnum*) and a destruction of the forests. Over the main oak forest areas hazel was probably the dominant shrub as it remains today. Both elm (probably wych elm) and the English lime (now comparatively rare as a wild species, see page 175) were widespread in the forests. During this Pollen Zone VII the land continued to sink till the sea level was some 25 feet or more above its present level (giving the Neolithic raised beach) though many of the submerged forests round our coasts, indicating a land level higher than at present, are of this age.

About the end of this period Britain was invaded by Neolithic peoples who used polished stone implements and introduced pottery. Far more important, they were the first inhabitants of this country to practise agriculture—but corn-growing on round or oval fields with the aid of hoes or deer-antler picks was subsidiary to grazing. Their flocks and herds began that great modification of the natural vegetation which was destined to go on continuously until the present day. They naturally settled first where there was relatively open ground or sparse woodland easily cleared, and where there was perhaps natural grazing for their sheep, goats and cattle. So their choice fell on the chalk downlands of Salisbury Plain, the South Downs and the Yorkshire

Wolds as well as on the Cotswolds and some of the Carboniferous Limestone uplands as in the Mendips and southern Pennines. Gradually they seem to have pushed back the forest and to have extended their grazing and cultivated lands. This extension would seem to have been encouraged by the climate becoming somewhat drier so that trees on the slopes suffered and there was a general dying back. Neolithic man mounted on a Celtic type pony and with hunting dogs doubtless hunted the wild animals of the woods but by this time there were few animals which are not members of our present fauna. Amongst the larger animals then existing which have been exterminated in historic times were the brown bear, wolf, beaver and wild boar. The Irish Elk, or Giant Irish Deer, may have survived in places into Neolithic times. There were certainly reindeer, roe and red deer (the fallow deer was apparently re-introduced in historic times). The great fierce auroch (*Bos primigenius*) was probably all too familiar and dreaded by Neolithic man. At least one, if not all, of the three species of hare still occurring in Britain was in existence, but there were no rabbits. There were shrews and field mice but no rats or mice to infest the early homes of Neolithic man. The earliest animal colonists following the retreat of the ice were able to find their way to Ireland and the western isles: later arrivals such as the mole, not found in Ireland, spread over Britain after the severance of Ireland. Britain has been cut off from continental Europe continuously since Atlantic times and it is largely since that time that various local races have evolved. The late Dr. J. C. Willis in expounding his Age and Area hypothesis for plants put forward the general concept that the smaller the area occupied by a species the younger in point of evolution it was likely to be. Though there are many exceptions, most distributions within Britain support this general hypothesis.

THE SUB-BOREAL PERIOD

It is clear that the mild moist climate of Atlantic times gradually became drier so that pine spread again over the eastern Fenlands and many Scottish and Irish bogs dried up. The downlands seem to have become somewhat too dry for cultivation so that the tiny fields, hand cultivated by Neolithic man, were gradually deserted by Bronze Age man who succeeded. Bronze Age burial places (tumuli) are numerous on the downs but actual habitations seem to have been on lower ground. Bronze implements date in Britain from about 2,000 B.C. to 500 B.C. Curwen believes that in the last part of this time Celtic immigrants introduced a light two-ox plough and established the Celtic field system of small rectangular fields (see Plate Ia). During this period the lime tree almost disappeared but beech and hornbeam began to spread. Forest extended to higher levels than at present—to 3,000 feet in Central Scotland.

THE SUB-ATLANTIC PERIOD

A cooler and damper climate seems to have set in about 500 B.C. Oak forest remained dominant over most of the lowlands but the pine retreated and there was active peat formation where forests had previously flourished so that bogs and fens expanded over both uplands and lowlands. The beginning of this period was marked by the settlement of the Iron Age peoples—Celtic invaders who used iron tools and weapons. Though there have been climatic fluctuations since, the Sub-Atlantic period saw the initiation in broad terms of the general type of climate we still enjoy. Manley considers there was a minor amelioration in Roman times, then a deterioration, again an improvement in the 6th to 8th centuries, followed by a wetter period about 1100 A.D. As Tansley has pointed out, none of these minor fluctuations has been sufficient to change the main features of vegetation, though they may have made the cultivation of warmth-loving plants such as the grape vine more successful than it would be today. We may accordingly use our experience of the present day to reconstruct the vegetation as it was when the Romans first saw Britain—or as it was when the later Iron Age invaders, the Belgae, settled here shortly before the first Roman expeditions. The position is summarized by Tansley in these words : "While there was fairly extensive cultivation on the chalk and also on the loam soils of the south-east, most of the English lowlands, for example the Weald and the Midland Plain, were covered with oak forest, mainly uninhabited and harbouring wolf, lynx, and bear, besides numerous deer. Oak forest also occupied the sides of the valleys in the hill and mountain regions of the west and north, giving place to pine and birch woods at higher levels and on the poorer and more sandy soils. In the extreme north of Scotland there was little or no oak even at low altitudes, though oak forest filled the bottoms and lined the lower slopes of the larger glens in the Central Highlands. In the limestone regions, for example on the Mountain limestone of the Northern and Southern Pennines, ashwood, the remains of which may still be seen in the Derbyshire dales, probably covered the sides of the valleys . . . . . Throughout the lowlands waterlogged areas were studded with many small meres surrounded by marsh or by fen, partly covered with willow and alderwood . . . . . the most extensive (now drained and cultivated) occupying the sites of old estuaries from which the sea retreated and the salt was washed out by rain. Towards the sea these estuarine fens pass into saltmarsh where the tides still have access. When the alkaline fen peat, fed by lime-containing ground water, has grown up above the winter water level, bog plants, largely bog-moss (*Sphagnum*), begin to replace the fen vegetation and form bog—or mosspeat which receives the water directly from the abundant rainfall of a wet climate and is kept moist by the damp air . . . such *raised bogs*, their surface slightly convex and above the level of the surrounding fen, remain in the central plain of Ireland, though in England most have been drained and destroyed. Bogs forming the same kind of peat occupy wide undrained areas not much above sea level in the wettest climates of all, i.e. in western Scotland and western Ireland, and these are called blanket bogs because they cover the country like a blanket. Blanket bogs also spread over high-lying plateaux where the drainage is poor and the rainfall heavy, as on Dartmoor, the Pennine plateaux, and in the Scottish Highlands. All these features of still existing fen and bog vegetation probably took shape at the beginning of the Sub-Atlantic period." (Tansley, 1941, p. 12).

Such was the country into which the Belgic peoples penetrated, occupying the southern chalklands and loamy soils from the Sussex and Kentish coastal plains, through the Thames Valley into Essex, Hertfordshire and East Anglia. The powerful little kingdoms founded by these peoples sometimes resisted, sometimes aided the Romans, but their agriculture was undoubtedly efficient enough to produce a surplus of corn and cattle for export.

We are concerned especially in this book with the effect of man's activities on the evolution of the landscape. Men of the older stone ages —the Eolithic, Palaeolithic and Mesolithic—have left their stone axes to be unearthed from gravel deposits but little if any sign of their existence remains on the landscape. From the arrival of Neolithic man in Atlantic times it is different. There are monuments, often conspicuous ones, in the landscape to remind us of the activities of successive waves of settlers.

Silent reminders of the Neolithic age are the megalithic tombs and long barrows concentrated on the Cotswolds and the chalk plateaus centred on Salisbury Plain, in Pembrokeshire and north-west Wales, with many also near the Scottish coasts and with outlying examples in the Wolds of Yorkshire and Lincolnshire. The great stone circles probably belong to the Early Bronze Age and the various relics of the Bronze Age as a whole make it clear that the population was concentrated on two types of country—the open downlands and along the more accessible or easily cleared valley lands. This is abundantly clear from the accompanying map where each dot represents a Bronze Age 'find'— the heavily forested clay lands were clearly avoided.

SUB-ATLANTIC PERIOD	Romano-British Period	A.D. 43—C. 400 A.D.
	Early Iron Age	
	Iron Age C (La Tene II and the Belgic settlers)	75 B.C.—40 A.D.
	Iron Age B (La Tene II)	250 B.C.—75 B.C.
500 B.C.	Iron Age A (Hallstatt—La Tene I)	500 B.C.—250 B.C.
	Bronze Age	
	Late Bronze Age invasions	1,000 B.C.—500 B.C.
SUB-BOREAL PERIOD	Middle Bronze Age	1,500 B.C.—1,000 B.C.
	Early Bronze Age invasions of "Beaker" peoples	2,000 B.C.—1,500 B.C.
	Neolithic Age	2,500 B.C.—2,000 B.C.

The earthworks or hill forts which conspicuously crown so many hills in Britain are often difficult to date but most are probably from the early Iron Age onwards. Those on the hill ridges of Lowland Britain command wide views : often they are in the midst of lands which show traces of the small rectangular Celtic fields. This points to the fact that when they were constructed the land around, if ever it had been forested, was already cleared. In the west many earthworks are of later date and may have been constructed against the Romans and even later invaders. Where they are found in foothill belts it is clear evidence that the forest had been cleared at least for considerable distances around. The greatest earthwork fortress in Britain—Maiden Castle in Dorset— is in the main Iron Age B though on a Neolithic site. An Iron Age date is usually assigned to the originals of the white horses cut into the chalk hills (see Plate II) and the giant man at Cerne Abbas in Dorset.

More difficult to date and explain are the lynchets found on hill-sides in chalklands and elsewhere. They appear to be narrow cultivation terraces perhaps resulting from ploughing parallel to the contours and turning the spit downhill. A fine series is shown in the air-view Plate Ib. It is difficult to visualize their deliberate construction by the light two-ox plough (or *aratrum*) which is believed to have been used until the Belgae introduced the much heavier *caruca*—with a broad-bladed share and a long, stout, concave coulter so set as to turn the furrow slice over. It may have been wheeled and was almost certainly

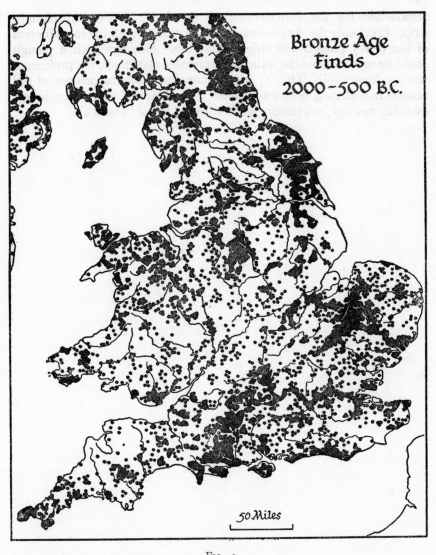

Fig. 3

Bronze Age Finds 2,000—500 B.C. (*after L. F. Chitty and Cyril Fox*)

This map suggests that in the Sub-Boreal period Bronze Age man had found
many of the most fertile lands—the Thames Valley terraces, the Sussex
Coastal plain and the Cambridgeshire lowlands

responsible for the introduction of the long, continental plough-strip. Previously the Anglo-Saxons were credited with the introduction of the heavy plough and strip cultivation. Although such a plough could be used to tame the hitherto intractable clay lands, the preference was for loamy soils. Thus in pre-Roman times the importance of soil texture had already been established. Two thousand years later, in our own day and age, soil texture is still the vital factor in land use.

CHAPTER 2

ROMAN RULE*

THE ROMANS brought to Britain national planning and thus anticipated many of the activities of the Coalition Government of
1941-45 and the Socialist Government of 1945-51 by about two thousand years. When Julius Caesar led his probing expedition of 55-54
B.C. he found his way inland from the Kentish coast barred by great
stretches of oak forest (the *horrida silva* of Tacitus) and he must have
realised that the control of such a country, inhabited by scattered or
wandering tribes at home in their woodland fastnesses, must depend
upon the establishment of well equipped ports of entry, good roads
into the trackless interior and fortified civil settlements at strategic
points, military posts on the frontiers.

Although it is the name of Julius Caesar that is associated in popular
imagination with the Roman invasion of Britain he was in fact only
concerned with transient military expeditions. True, he left descriptions
of Britain based on his own observations, and provided material for the
accounts written by Diodorus Siculus about 45 B.C. and by Strabo
about 7 B.C. The conquest of Britain did not take place for nearly a century after Caesar's expeditions ; the main invasion—planned by the
Emperor Claudius—was in A.D. 43. Within the next four years the
greater part of south-eastern England came under Roman domination,
the temporary frontier following roughly the line of the Fosse Way from
Lincoln to Cirencester and Exeter. The Civil Zone, or zone of civilian
occupation, was steadily expanded until it came to coincide very closely
with what we know today as Lowland Britain, whereas the hilly or
mountainous western and northern country was occupied almost entirely for military purposes and constituted the Military Zone. The
windswept moor-covered uplands of Exmoor and Dartmoor did not
attract the Romans but doubtless the mineral riches of Cornwall did, so

* I am greatly indebted to my good friend Professor Eva G. R. Taylor for valued
comments on chapters 2 to 7, and to my colleague Professor E. M. Carus-Wilson for
helpful criticism of chapters 2 to 5 and 9.

that the south-west beyond the gateway city of Exeter was partly an area of civilian settlement. Agricola completed the conquest of Wales in A.D. 78 and of northern England in the following year—indeed he extended the boundary of Roman Britain in A.D. 80 to the narrow waist of Scotland between the Firth and Clyde. But the Picts gave continued opposition and so the Romans withdrew to the line of the Tyne-Solway. There, from Wallsend on the Tyne to Bowness-on-Solway over a total length of 73 miles Hadrian's Wall was constructed in A.D. 126. In those places where the wall takes advantage of the vertical scarp of the Whin Sill it is still possible to stand where the Roman sentries stood and look northwards over the rolling ground whence the relentless war-like Picts were ever liable to strike. Life on this outpost of the Roman Empire has become familiar not unfortunately through contemporary writings but through the vigorous prose and imaginative genius of Rudyard Kipling. Kipling merely in fact translates what he knew of the life of the British Tommy on the outposts of Empire in India to some eighteen centuries earlier and makes it the life of the Roman legionary on the outposts of Empire in northern Britain. Perhaps the parallel is even closer than Kipling imagined when he wrote.

After the construction of Hadrian's wall a determined effort was made to extend the bounds of Empire and the Antonine Wall, 37 miles in length, was built along the Forth-Clyde line in A.D. 143. But the northern extension did not last long—indeed between the years 180 and 185 the Romans lost control of both lines of defence. When peace was restored Hadrian's Wall became the recognized northern limit of Roman Britain and so remained for nearly two hundred years till it was finally abandoned in A.D. 383.

The two outstanding questions are : what sort of a country did the Romans find when they invaded and conquered Britain; second, what were the permanent marks they left to indicate their three and a half centuries of occupation ?

There is abundant evidence that the land had been sinking relative to the sea in Neolithic times and that many former valleys had been invaded by marine waters giving rise to numerous sheltered if shallow harbours. The main subsidence of the coasts lasted from about 3000 B.C. to 1600 B.C. By Roman times many of these inlets were—though the Romans probably did not know it—already silting up rapidly. This process has gone on more or less continuously ever since despite some renewed subsidence, and the sea inlets of Roman times are now land— shown as occupied by alluvium on the geological map but in many areas

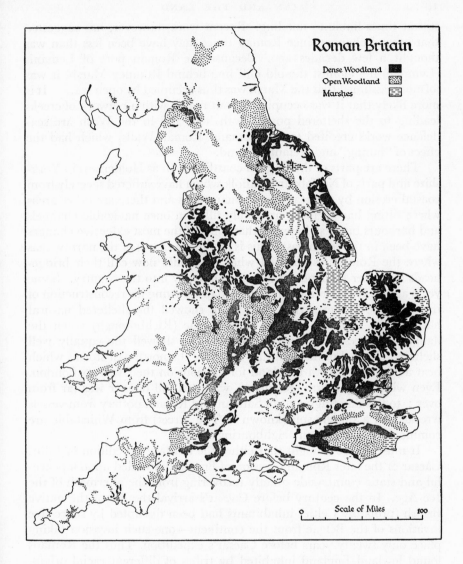

Roman Britain

Dense Woodland

Open Woodland

Marshes

Scale of Miles

0   100

FIG. 4

The Forest Cover of Britain in Roman Times as shown on the map of Roman
Britain published by the Ordnance Survey

This map is now generally regarded as showing a minimum of forest coverage

still at times liable to flooding. Recent work, however, has suggested that coastal changes since Roman times may have been less than was thought a few decades ago. Because the Roman port of Lemanis (Lympne) lies against the old cliff line behind Romney Marsh, it was formerly assumed that the Marsh was thus occupied by open sea. It is more likely that it was occupied by salt marsh with a network of creeks leading to the sheltered ports. Both here and in the Wash are sea-defence works credited to the Romans (Roman Walls) which had the effect of "inning" and reclaiming land.

There are parts of the English coastline such as Holderness in York-shire and parts of Norfolk and Suffolk which have suffered severely from coastal erosion by the sea since Roman times and there are other areas where silting has partially or wholly filled in once navigable channels and harbours but it so happens that some of the most extensive changes have been in the south-east. This is along the shores of the narrow seas where the Romans established what we should now call their bridge-heads and later their permanent ports of entry into the country. Some years ago T. Rice Holmes published maps showing his reconstruction of the coastline of the period. His maps showed the sheltered natural harbour the Romans enjoyed at Rutupiae (Richborough) when the dangers of egress had been overcome : he showed the equally well sheltered channel of the Wansum avoiding the north-east storms which beat against the North Foreland for galleons on their way to London. Even when Regulbium (Reculver) was past, all danger was far from over : to this day fragments of Samian and Roman pottery from vessels wrecked on a shingle spit (known as The Street) from Whitstable are commonly washed up on neighbouring shores.

It must not be thought that either the military expedition of Julius Caesar or the later Roman invasion of Britain descended upon a peace-ful and static countryside quietly recovering from the aftermath of the Ice Age. In the century before Caesar's arrival the life of the native British (probably Celtic) inhabitants had been disrupted by successive invasions of the Belgae from the continent—one such invasion taking place only twenty years before Caesar's expedition. Thus the Romans found lowland England inhabited by tribes of different racial origin, probably at very different cultural levels and at enmity one with another. It is not surprising that the Romans found some tribes ready and willing to cooperate, others fiercely opposing the invaders. These varied peoples of the lowlands were in turn distinct from the Celtic peoples of the mountainous west and north.

Modern aerial photography and the expert interpretation of O. G. S. Crawford have revealed the widespread cultivation practised by these pre-Belgic and pre-Roman inhabitants of Britain. It is not surprising that the groups of fields are most commonly found in the uplands such as the chalk downs (Plate Ia) where the natural forest if it existed at all was sparse and more easily cleared than the dense oak woodland of the valleys. For precisely similar reasons many of the 17th century settlements in New England were established by the Pilgrim Fathers on the higher ground whilst valleys since settled were neglected. If one may draw conclusions from the under-developed lands in the world today it is highly probable that land was cleared and the growth of grass encouraged by the use of fire as well as by the action of grazing animals discussed later. The extent to which firing natural vegetation has been used all over the world is rarely appreciated. As already stated these ancient Britons grew enough grain for their own use and had a surplus for sale when Roman armies created a demand or means of export were opened up. They were sheep farmers and cattle-keepers too : it would seem that since Neolithic times they had been experts in the construction by using puddled clays of ponds on the downs where running water is absent and natural ponds are scarce*. There are those who consider that the situation of the pre-Roman settlements suggests a damper climate than at present but the relative ease of land clearance is a more likely explanation. The name 'dew-ponds' in the book quoted above begs the question : the ponds seem rather to have been designed to collect surface rainwater.

It may well be that a clearer distinction should be drawn between Neolithic or pre-Celtic man in Britain, the Celtic peoples who invaded the country from about 1000 B.C. onwards, and the Belgic peoples. It is claimed for example that the Celts introduced the fundamental ideas of settled farming with stockaded yards including the farmstead, barns and byres, and that they used a primitive plough. Bronze and later iron (450 B.C. onwards) enabled the Celts to extend and intensify their farming. It may well be the Belgae who, being of Celtic-Germanic origin and skilled in the use of iron, introduced the heavy eight-ox plough complete in essentials with coulter, share and mouldboard, which was destined to leave such permanent marks over Britain for more than a thousand years. Possibly the Romans found already developed, side by side, the Celtic square field cultivation and the Belgic strip farming and that it was the latter which yielded a grain surplus.

* Hubbard, A. J. & G., Neolithic Dew-Ponds and Cattle-Ways 1905 and 1916.

It would seem that each of the tribes of pre-Roman Britain occupied a loosely defined area and probably each had a main settlement or capital. In renaming and redeveloping some of those capitals, the Romans perpetuated the tribal name, thus Cirencester was Corinium Dobunorum, Caerwent was Venta Silurum.

The Romans imposed upon rural tribal Britain an urban industrial civilization. The Roman Empire was based upon the imperial city of Rome : the Roman leaders were city dwellers who were fully convinced of the virtues and benefits of an urban life and they set out to confer these supposed advantages upon the provinces of the Empire. They believed firmly that they had the 'know-how' two thousand years before Europe and America approached the underdeveloped lands with the same idea.

The general pattern of the Roman development of the backward or undeveloped province of Britain has been repeated again and again in the history of the world. The first century or so was occupied mainly in the extension and military consolidation of the province. A first need was the establishment and development of good ports of entry from the continent. At the narrowest crossing point of the surrounding seas Dover (Dubris) had a tiny natural inlet and, more important, one easily defended by forts on the towering chalk cliffs on either side which at the same time could guard the easy road up the valley to the natural route-way along the dry open chalk downs towards London. It is little wonder that the white chalk cliffs of Dover stood, for the Romans, as symbolical of Britain just as they did in later years to the French for Albion, and later still have inspired nostalgia even in hardened world tourists.

It would seem that Dover was the principal military port whereas the port for civilian use was Richborough (Rutupiae), then at the sheltered entrance to the Wansum channel, now some distance from the mouth of the Stour, the entrance to which is sheltered by the treacherous but useful Goodwin Sands—which may have been an island in Roman times. Richborough came into use again as a port during the First World War. As a young Engineer officer in charge of heavy bridging at the Base Depot at Calais it was often my duty to consign laden rail trucks by railway ferry to Richborough whence they could be sent direct to certain destinations by the railways of south-east England—except where tunnels and bridges were too narrow. I reflected at times how some seventeen or eighteen centuries earlier some Roman officer at what is now Boulogne had similarly consigned his goods to the self-same port of Richborough.

Aerofilms Library

Plate Ia.—Traces of Small Square " Celtic " Fields, Saxon Down, Sussex (*p. 9*)

b.—Lynchets near Chippenham, Wiltshire (*p. 12*)

Aerofilms Library

Plate II.—The White Horse at Westbury, Wiltshire

The horse has been made by cutting the grass turf away from the steep scarp slope and exposing the white chalk. Above the horse is a ploughed field (the soil is shallow and the plough turns up white chalk) inside an Iron Age fort. To the left the steep slope leads to the fertile belt with cornfields on chalk marl. In the top left corner a close county of smaller fields with hedges and hedgerow trees marks the transition to clay soils. The village is sited on a spring line (*see pages 12 and 34*)

Much trans-channel traffic passed through the Wansum Channel, past Reculver (Regulbium) and up the Thames. It is a little more difficult to assess the importance or indeed to decide the exact position of Portus Lemanis, but presumably it was below the old cliff line at Lympne along what is now Romney Marsh, replacing as suggested above the former salt marshes. There is certainly a Roman road from Lympne to Canterbury. Another port guarded the inlets near Pevensey. A Roman road, later called Stane Street, led to London from the neighbourhood of Chichester and doubtless from a port somewhere on one of the sheltered inlets in that vicinity.

Porchester guarded the network of channels near the modern Portsmouth. A small Roman port existed near the site of Southampton but the sheltered haven of Southampton Water was too far west to be greatly used by the Romans. The Roman fort at Carisbooke Castle, Isle of Wight, may have been connected with the defence of this entry.

On the Essex coast evidence of Roman ports is less clear but doubtless the great centre of Colchester was specially served. Sites of forts have been traced at Brancaster (Branodunum), Burgh Castle (Gariannonum), Walton Castle (which was south of the river Deben near the modern Felixstowe and finally disappeared into the North Sea in the mid-eighteenth century) and Bradwell (Othona). All the ports were defended by forts and the defence of the "Saxon shore" played a large part in Roman strategy during the later years of the occupation. Purely military forts existed elsewhere from Yorkshire in the north to Cardiff and Holyhead in the west.

From their ports on the south-eastern coasts the Romans placed great reliance on good roads inland. They built roads to serve both military and economic uses just as nineteenth century engineers built railways. The idea of using broken stone to form a hard surface was not understood until the eighteenth century and the Romans laboriously paved their roads with large flat stones. It is interesting that the Romans have remained great road builders—as the work of the Italians in East Africa and Ethiopia in the present century has shown. There are clearly points of comparison between the work of the Romans in Britain and the British in the nineteenth century in India or perhaps better in East Africa—the good port at Mombasa: the railway inland from which control could be exercised, and the country opened up to trade. By force of necessity—dictated in other words by geographical conditions—the Roman roads converged on London. Whatever there may have been in the way of pre-Roman settlement at London, its genesis as a great city dates from Roman times. Where two gravel-capped flat hills gave firm flood-free ground overlooking the river was the ideal site. The one hill is that now crowned by St. Paul's Cathedral, the other (from which it was separated by Wall brook) by the Royal Exchange and Leadenhall Market. Whilst there is no evidence that London was ever regarded by the Romans as the administrative capital of Britain it quickly became the commercial capital mainly by reason of its focal position. As early as A.D. 60 it was celebrated as the home of the merchants who imported oil and wine from Italy and pottery from Gaul, and exported corn and slaves. It had already developed such manufactures as glass and pottery making.

As the Romans extended their control over the country they followed a familiar pattern. Where existing settlements suited their purpose and were well-sited they took them over and Romanized them. Thus the Belgae who preceded the Romans had established St. Albans (Verulamium) and probably Winchester (Venta Belgarum) and had taken over and enlarged the old British capital of Colchester (Camulodunum). Elsewhere the Romans used the chief settlements of British tribes as the full names of the towns often indicate. This is clear from Caistor-by-Norwich (Venta Icenorum), Canterbury (Durovernum Cantiacorum), Exeter (Isca Dumnoniorum) and many others shown on the map (Fig. 5). Sometimes the Romans chose a site a short distance from the existing settlement, as when the site selected for Colonia Victricensus was a short distance from Colchester. In fact the Roman Verulamium was not on the exact site of the earlier town.

In other cases the Romans selected, with great geographical skill, new sites. Thus Rochester (Durobrivae) was clearly designed to guard the vital crossing of the Medway on Watling Street between Dover and London and in the same category we may put the three great garrison towns or legionary fortresses of York (Eboracum), Chester (Deva) and Caerleon (Isca) from which were serviced the hundred or more forts established as strategic points in Wales, the north of England and the Southern Uplands of Scotland. There is an obvious parallel with the British on the North-West Frontier of India, with their bases at Peshawar and Rawalpindi.

The siting of Chester affords an interesting example of Roman appreciation of site factors. It stands on the north bank of the River Dee where a knoll of red sandstone rises a hundred feet above sea level and above the surrounding wet glacial clays. In Roman times the estuary of the Dee must have been a wide open arm of the sea and Chester stands

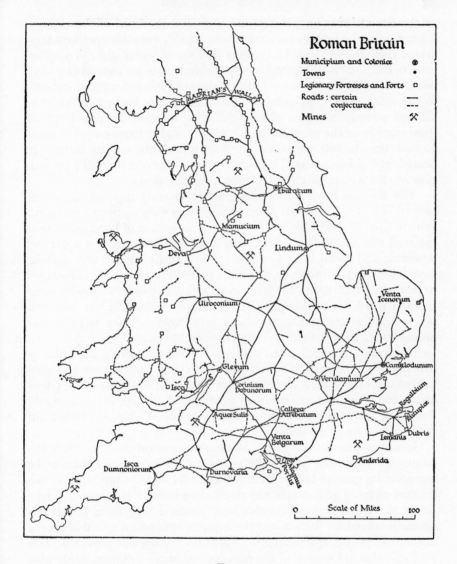

FIG. 5
The Roads and Towns of Roman Britain
Spellings used on this map are those of the Ordnance Survey. Mamucium
=Mancunium; Uiroconium=Uriconium; Eburacum=Eboracum.

at the lowest bridging point of the river and the head of the estuary. Silting up of the estuary occurred more rapidly here than perhaps anywhere else around the British coasts and has obscured the old significance of the site. Well located in its immediate site, the fortress city of 56 acres controlled the route to the lead mines of Flint and along the North Wales coast to the copper mines of Anglesey. The legions stationed in Chester served the chain of forts guarding lowland Roman England from raids from the mountains of Wales. Notable amongst other great Roman foundations which have grown today into mighty towns are Manchester (Mancunium), Newcastle-upon-Tyne and Cardiff : in each case was the sure appreciation of favourable site factors.

The Roman town followed a pattern characteristic of an urban people. The walls enclosed a rectangular area varying from a few acres to a hundred or more—reaching 330 acres in the case of London. Within the wall was a grid-iron street plan with the Forum as the centre of economic and social life. The concept of a 'civic centre' was nothing new to the Romans: town halls, public baths and other urban amenities were an essential part of the Roman system of urban life. A prominent position was likewise occupied by the temple or temples. It has been suggested that the public buildings in a Roman town were out of proportion to both the numbers and needs of the people.

Naturally the first Roman towns in Britain were primarily military strongholds : the earliest development of civilian settlement was probably the establishment of the four *coloniae* of Colchester, Lincoln, Gloucester and York as settlements of time-served Roman soldiers. Only these four towns and the solitary *municipium* of St. Albans carried full rights of Roman citizenship.

It was not until later in the Roman occupation that one finds the establishment of *villae* outside the towns and the growth of what may be considered a class of landed gentry. The *villa* was the precursor of the country estate. The Romans had no illusions about the British climate : under the house was a chamber and 'boiler room' from which hot air ducts passed up the walls of the living room and an effective system of central heating, and incidentally of grain drying, was provided. It is probable that the owner of the *villa* cultivated the adjacent fields with the aid of slave labour but may have let off part of his estate to tenants or to serfs. Probably the owners of *villae* marketed their produce in the towns for most are situated within easy reach of urban markets.

Away from the Roman towns, *villae*, and roads rural Britain existed much as it had done from earlier times. Although under Belgian and

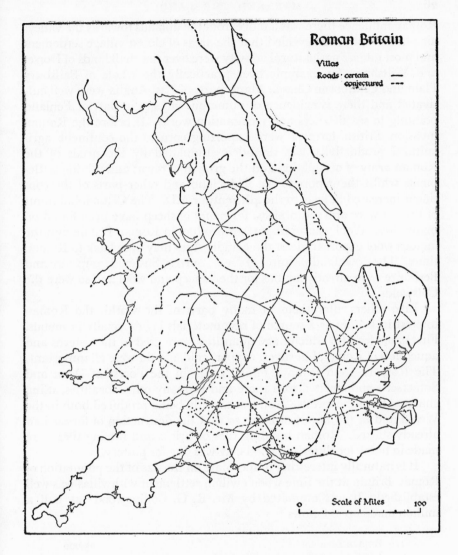

Fig. 6
Country estates of later Roman times
The Roman or Romano-British landed gentry who built themselves 'villae'
or country houses rarely sited them far from a main road

Roman influence there was a move from the uplands towards the valleys air photography has revealed that the areas of closest village settlement and most intense agricultural activity were over the chalklands of Dorset and Wiltshire—for example over practically the whole of Salisbury Plain and Cranborne Chase. Many parts of East Anglia were well cultivated and there is evidence of numerous villages in part of Fenland pointing to sea defences and reclamation works. If before the Roman invasion British farmers were exporting corn to the continent agricultural productivity was doubtless stimulated by the needs of the Roman army of occupation and the growth of towns and mining settlements whilst the export to the Rhineland and other parts of the continent increased at least to the 4th century A.D. The Celtic inhabitants of Britain were also pastoralists, for bones of sheep have been found on many sites. Wool was an important product in Roman Britain and the famous wool export trade of the Middle Ages may date back to Roman times. The downland pastures of Kent, Surrey, Sussex, Hampshire and Berkshire were already famous for their sheep and wool and so were the Cotswolds.

Like others after them in many parts of the world, the Roman interest in Britain as a source of raw materials lay especially in metals. First came lead (of which huge quantities were used in the cisterns and aqueducts of Rome), but silver, copper, gold and tin were all important. The Romans seem to have mined coal—in the Forest of Dean and Somerset—and smelted iron there and in many other localities, using charcoal from the forests. Iron was already being produced both in the Weald and in the Forest of Dean so that the destruction of forests had already started. Roman building tiles or bricks and roofing tiles were made in many localities and good clay was dug for pottery.

It is naturally interesting to speculate on the size of the population of Roman Britain at the time when civilian settlement with villas was well established. It was estimated by Mr. R. G. Collingwood* at half a million made up as follows :

1.	Roman London	25,000
2.	Large towns—20 with 5,000 each	100,000
3.	75 smaller towns, 1,000 each (50 towns actually known)	75,000
4.	1,500 villages with 100 each (700 villages known)	150,000
5.	1,000 villas each 50 persons (500 villas known)	50,000
6.	The Army and dependants	100,000
		500,000

* Town and Country in Roman Britain. *Antiquity*, III, 1929, 261-76

Sir Mortimer Wheeler† criticized this estimate as too small, considering that the military zones of the north and west and the mining areas had been overlooked. There is doubt too whether the estimated 150,000 rural inhabitants could have produced the volume of food sufficient to feed themselves, the towns, the Army of occupation and a surplus for export. Sir Mortimer Wheeler preferred an estimate of one and a half millions.

Taking the world as a whole at the present day about 3,500,000,000 acres of arable land exist to supply the needs of 3,250,000,000 people. Allowing food alone it takes the produce of about an acre to feed one person. This is true to the intensive cultivation of north-western Europe coupled there with a high output per acre and a high standard of living: it is true of the underdeveloped areas with primitive forms of agriculture coupled with a low output and lower standard of living. With the much lower yields then obtained we may hazard a guess that it needed the produce of at least two acres probably three (with no allowance for fallow) to support each person in Roman Britain and we know there was an export. The area cultivated and cropped cannot have been less than a million acres, and may well have been nearer 4,000,000. In addition were the sheep and cattle pastures. These figures we may apply to lowland England alone: too little is known of Wales and Scotland. Already then some 10 to 20 per cent. of the surface of England was cultivated: already some of the most fertile soils and the best climatic conditions were known: already Britain had established a reputation for its wool and was famed for the number of its sheep. Although several writers have referred to the large areas 'under the plough', a considerable part of land cultivation was probably still by hand.

Sir William Ashley in his well-known book, *The Bread of our Forefathers*, has shown how difficult it is to be certain even what were the bread grains of our ancestors. Whilst *frumentum* (cf. modern French *froment*), *siligo* (cf. *seigle*), *hordeum* and *avena* were used so long as records were in Latin, i.e. to the eighteenth century for wheat, rye, barley, and oats respectively, the first may often mean simply corn, as the British now use the word, or small grains in the American sense. *Triticum* is specifically wheat. On the whole the evidence is that Britain in Roman times was growing wheat as the bread grain and that rye was introduced by the Anglo-Saxons and that thenceforth it was in general mainly wheat for the rich and rye for the poor. Even then, as for the next thousand years, barley and oats—perhaps with peas, beans and acorns—

† R. E. M. Wheeler, *Antiquity*, IV, 1930, 91-5.

entered into the national loaf of the poor. Reference will be made in later chapters to plants and animals which may have been introduced by the Romans—including possibly the chestnut, the walnut and the vine. Amongst the Romans the horse was a badge of rank but the pre-Roman inhabitants of Britain were themselves already using horses.

Before the Romans left Britain, most of our principal towns were established, many of our existing main roads already constructed. Few of the Roman towns have failed to survive—the chief being Silchester (Berkshire) and Uriconium (near Shrewsbury)—and most of the roads, though not all, have survived into the age of motors as main highways. As far at least as the south and east are concerned most of the main ports had been established : Roman London was already a city of world note. The period of Roman rule did not of course come to a sudden end. What has been called the 'last phase' stretched from 410 to 582 A.D. when Romano-British or Romanized British forces struggled against insurgents at home and invaders from without. After the Picts and Scots broke through Hadrian's Wall, the whole position steadily deteriorated. Roman farms or villas were vulnerable units, Romans and Romanized Britons withdrew to the towns. All Roman troops and officials were withdrawn in 410 or soon after. A notable victory over the Romano-British in 457, and their retreat to London may be regarded as marking the beginning of the end. Nevertheless about 500 A.D. a warrior who has been identified as Tennyson's King Arthur secured a measure of ordered control until his death in 550, when the old Roman society finally disintegrated.

Plate III.—Anglo-Saxon Strips at Napton-on-the-Hill, Warwickshire

This air view was taken in the late afternoon and the small hedgerow trees cast long
shadows over the ridged fields once ploughed and cropped, now grassed (*p. 34*)

Plate IV.—A Typical Welsh Valley in Montgomeryshire, Llangynog

A mosaic of hedged fields occupies the valley and the fields climb the slopes (except where they are too steep) to the economic limit of cultivation or improvement. The slopes above were probably once forested: now they are occupied by grass moors with darker patches denoting bracken

CHAPTER 3

SEVEN HUNDRED YEARS

SEVEN HUNDRED YEARS elapsed between the main withdrawal of the Roman legions in A.D. 383 and the Norman conquest in A.D. 1066—or to take a more important date in the history of land use, the Domesday Book of 1086. It has proved extremely difficult to reconstruct the history of those seven centuries not only because of the scarcity, even the non-existence, of written records, but also because the Roman construction in brick and stone gave place to the use of wood with resulting destruction without trace of the buildings of the period. Building in stone, and then only for castles, churches and cathedrals, was not revived until Norman times—apart from a few Saxon churches antedating the conquest by a few years. Brick and stone did not become general for other buildings until much later. It has truthfully been said that Roman London had more permanent buildings in brick and stone than the city was destined to see again until after the Great Fire of 1666.

Yet in those dark ages the humanized landscape of Britain took on much of its modern aspect. In particular the essentially English patchwork of arable, meadow, rough pasture and woodland emerged in a form which can be compared in general terms, except for the unenclosed arable lands, with the present day. In those centuries most of the villages and non-industrial towns which we have today had already been established : the majority of our rural parishes as well as our counties date from those times. Before the Normans arrived nearly all the good farmlands had, by trial and error, been found and used: if the term marginal land was not used its meaning was familiar enough. By Norman times relative land values of farmlands had been established according to inherent productivity and showing a pattern in most areas closely comparable with that of the present day.

Much of the story of the seven centuries is enshrined in the place-names of the countryside and an immense debt is owed to the patient researches of the members of the Place Name Society and the county

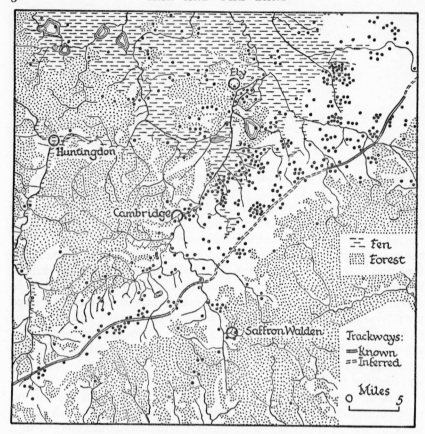

Fig. 7
Bronze Age Finds in the Cambridge Region (*after Cyril Fox, Archaeology of the Cambridge Region, 1923*)

At this time population was concentrated on the cleared lowlands bordering the fens and the fertile well-drained belt along the main Icknield Way. The clay lands were forested and devoid of settlers

volumes which the Society has published. Perhaps there has grown up as a result too great reliance on place-names. We recall that the Roman settlements are no longer known by the names by which they were known to the Romans. True, their present names by such suffixes as —chester (*castrum*, a fort or castle) recall their Roman origin but the fact remains that the names have been changed. How many of the towns and villages with Anglo-Saxon or Scandinavian names were in fact

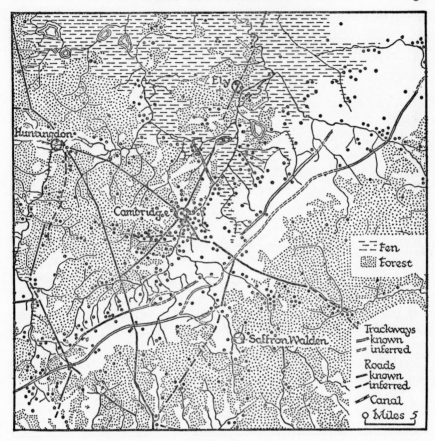

FIG. 8
Roman Remains in the Cambridge Region (*after Cyril Fox*)
Settlement was related to the Roman roads but there were many clearings in
the forest

earlier settlements taken over and renamed by the invaders ? This is
something which we shall probably never know. If we take the evidence
of place names, then the bulk of the settlements in Lowland Britain are
of Anglo-Saxon or Scandinavian origin. They are certainly not later.

The Romans had come to Britain as Empire builders, intent on adding yet another province to their world empire, convinced that the
Roman way of life was the right way for the world and one which, for
the benefit of conquered and conqueror alike, should be imposed upon

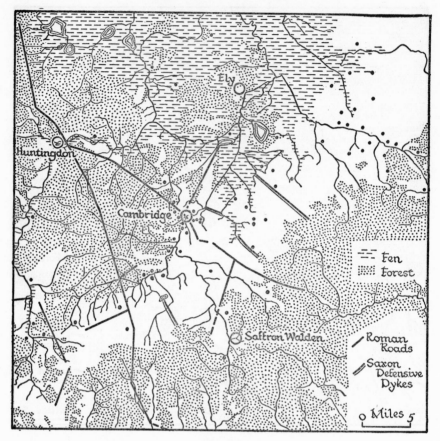

Fig. 9

Early Anglo-Saxon (Pagan) settlements in the Cambridge Region
(after Cyril Fox)

This map affords clear evidence of penetration by rivers and valley ways

all. Whilst the Romans left a lasting legacy of cities and towns with linking highways their influence on the scenery of the countryside as a whole was small. They left the British tribesmen and cultivators very much to follow their own paths. By way of contrast the Anglo-Saxons came as settlers, to possess the land and there to make homes for themselves and their descendants. Some historians would separate an early phase of conquest from a later one of settlement, but it would seem more likely that the two went side by side. Perhaps the closest parallel is with

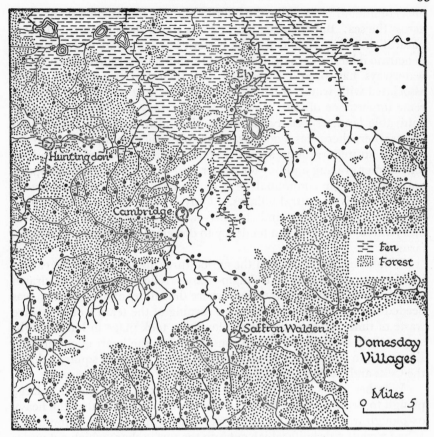

FIG. 10
Domesday Villages in the Cambridge Region (*after Cyril Fox*)
By late Anglo-Saxon times there were settlements over the whole region
except the undrained fens and where surface water is absent

the seventeenth century settlement of the Pilgrim Fathers in New England : there were battles and skirmishes with the American Indians
when they presented opposition, but much friendly contact. The New
Englanders acted in groups, established their settlements on a communal basis and laboriously cleared the land of wood to make way for
fields. This is precisely what their forefathers the Anglo-Saxons had
done in England a thousand years earlier. They had come in village
groups, established a settlement, cleared the land and then cultivated it

M.L.—D

on a communal basis. Like the Pilgrim Fathers, the Anglo-Saxons came across the sea. But their sea journey was short, their boats small. It is natural that they should use the eastern-flowing rivers of England to penetrate inland. Perhaps the extent to which they relied on navigable waterways has been exaggerated but their settlements are always associated with water. Like all good farmers they understood the supreme importance of both land and water : they sought out the fertile cultivable land, especially on loams, they always placed their settlements where there was a good water supply either from a surface stream or from an underground supply such as a spring. Waterside meadows for summer grazing and hay for winter feed were of the greatest importance also. Their communal form of life resulted in nucleated settlements—that is in central villages. Only on the clay lands with abundance of surface ponds and streams—where water in other words was everywhere—do we find a tendency to dispersed settlements or isolated farms.

Whatever may have been the agricultural implements of the earlier inhabitants of Britain, the Anglo-Saxons relied upon the ox-drawn plough. Especially on heavy land where the full team of eight oxen was needed to pull the plough, frequent turning of the team would involve waste of time and energy and so ploughing took place in long straight strips. Strip cultivation, still so conspicuous in the heart of continental Europe from which the Saxons came, replaced the small square fields of the Celts and has left an indelible impression on British scenery.

Whether or not the Celtic or Belgic peoples introduced a large plough to Britain in pre-Roman times there is no doubt that the heavy ox-plough was the main agricultural implement of Anglo-Saxon times. It had a coulter or knife which cut into the turf, a share which cut under it and a mouldboard which turned the sod over. The farmers ploughed so that they turned the sods in towards the centre of the strip, or *selion*. The breadth of the selion approximated to $5\frac{1}{2}$ yards or $16\frac{1}{2}$ feet, which is the modern rod, pole or perch. Indeed the Anglo-Saxon plough strips have been perpetuated in the strange British system of land measures. A rod or perch is a quarter of the modern chain and whilst blocks of selions or strips were known as furlongs or in the Danish area as 'longs', the commonly accepted derivation of furlong is a 'furrow long'. A furlong has been standardized at 220 yards, or an eighth of a mile. A parcel of land one furlong in length (or 220 yards) and a chain or 4 perches wide (22 yards) constitutes an area of 4,840 square yards or one acre. This was regarded as the approximate extent on average loamy land of a day's ploughing. The larger unit of land was the *hide,* an indefinite measure ranging from 40 to 120 acres, which may have been the area normally cultivable by an ox-team in a year. A hide, whatever its size, was definitely the land of one family—the *terra unius familiae* ot Bede. Interestingly enough the 1941 Farm Survey revealed for the first time that the average English full time farm holding which would now be farmed by the farmer, his family and one hired man approximates to 100 acres of crops and grass. In the same way the modern American farm averages 85 acres of tilled land out of a total holding with woodland and rough grazing which is often 160 acres. The hide, in just the same way, seems to have been the unit which supported a farmer of substance ; the smaller man or *gebur* of the Saxons had to be content with a yardland or virgate (which was a quarter of a hide) or a half virgate, say 15 acres, and which could be farmed with a pair of oxen. The true smallholder or *kotsetla* had to be content with 5 acres or so, cultivated by hand. He was only a part-time farmer with other duties to perform. In the Danish area of settlement in East Anglia—the Danelaw—a different terminology was employed : the hide being replaced by the ploughland or carucate which was definitely the area cultivable by the eight-ox plough team in a year. A ploughland was divided into eight ox-gangs or bovates.

It is interesting and not perhaps idle to speculate on the results of equating a hide or a ploughland with a balanced farm unit. The thegn (or thayne) of Anglo-Saxon times, replaced by the Lord of the Manor in Norman times, would be the large landowner with several hides or farm units. On average land the present day farmer needs 100 or 120 acres : where the land is of first quality as in the Fens or Vale of Evesham much less suffices ; where the land is poor and light much more may be needed. It may be that here we see the reason for the original range in size of a "hide" or "yardland" though later it may have become just a unit in taxation assessment. Similarly in mediæval and later times a farmer could plough with a single horse or pair of horses a far larger area of light land than he could if his farm were on heavy clay or "four-horse" land needing four horses to pull the plough. All these farm sizes ignore "rough grazing" (the old "waste") just as the American size quoted above ignores the woodland and unploughed land. It may well be that the present-day full-time farm-holding is one of the oldest units of the countryside.*

*Lord Rennell of Rodd has shown that the small manors of north-western Herefordshire had each two arable fields of 35 acres each and that boundaries have often remained almost unchanged for a thousand years.

This is not incompatible with the fact that over most of Anglo-Saxon Britain the village ploughlands were held in common, each man responsible for the strips which fell to his lot in the annual allocation, though the ploughing was done by the ox-teams shared by the village. There are farms today which are almost as fragmented as in the days of isolated strips in common fields. No fences separated the old strips though sometimes a baulk or slight earthbank marked the limit between one and another. In a few parts of the country long narrow fields at the present day indicate where enclosure by hedges followed the lines of the old strips and fields are, in fact, fossilized strips.

The normal practice in Lowland England except in the east and south-east was for each village to have three common fields—one in bread corn or rye, later wheat ; one in drink corn or barley and oats ; the third lying fallow. In some areas two common fields, one in cultivation the other lying fallow seem to have been the rule, and some historians would say this was the common system. We need not here enter into the controversy as to whether the two-field system was the older and gave place to the three-field so as to economize in land, or whether the two were contemporary and dictated by local physical conditions or fashion : suffice it to say both existed.

The earlier phase of the main Anglo-Saxon invasion, infiltration and settlement, before the conversion of the people to Christianity, was roughly the two centuries from 450 to 650, the Pagan Saxon period. The people came from the continent : their earlier settlements and their first kingdoms were established near the south and east coasts. It was natural that they should enter by the waterways emptying into the shallow but protected harbours which the Romans had used. This is well seen from the disposition of the early settlements in the Cambridge region as studied by Miss Chitty and Sir Cyril Fox (Fig. 9). When their boats, which probably drew as much as three feet when laden, could no longer pass it is natural that the newcomers should continue along the river valleys, since there side by side were the valley gravel soils or flood plain alluvium which they recognized as eminently cultivable, and the all-important water supply.

As experienced farmers the Anglo-Saxons had little use for the thin reddish and blackish soils of the chalk downlands which the Celts had used : the Anglo-Saxon ploughs would have turned up the chalk from below. At the same time they were repelled by the stiff deep clays of many of the vales and lowlands still covered with dense oak forest. Likewise they recognized the coarse sandy soils and plateau gravels as

Plate 1. THE GOLDEN VALLEY OF HEREFORDSHIRE

Typical of the man-made landscape of rural Britain, sheep and Hereford cattle grazing in the rich pastures of the valley floor, crops ripening on the well-drained hill slopes with moorland rough grazing above. The field divisions are planted hedges

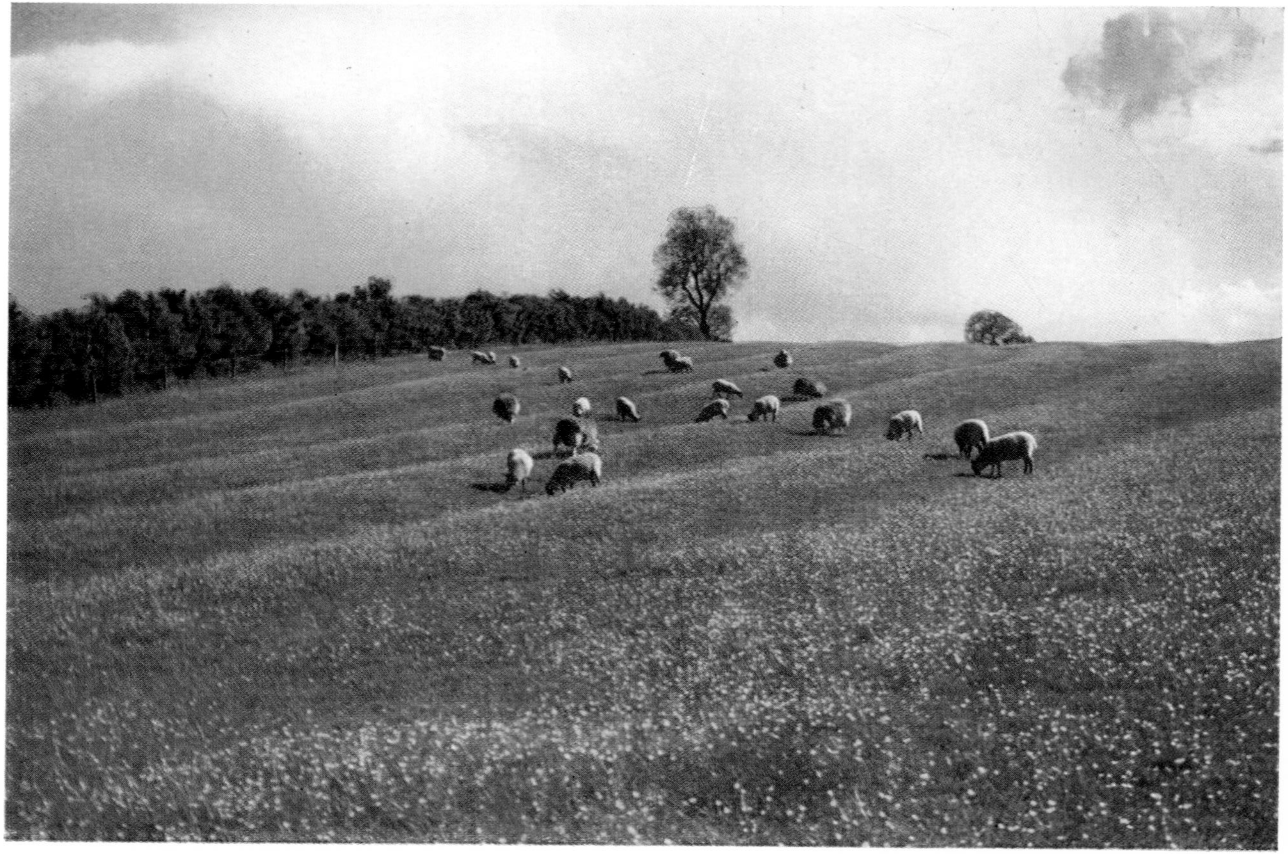

P. W. Bryan

Plate 2. RIDGED FIELDS, TILTON, LEICESTERSHIRE

The influence of Anglo-Saxon farmers is still seen twelve or thirteen centuries later. (p. 24)

Cultural Zones
of
Anglo-Saxon
Britain

Viking Settlements
in Western Britain

Anglian

Anglian

Anglo-
Saxon

Saxon

Jutish

Jutish

Miles
0   100

Fig. 11

The Cultural Zones of Anglo-Saxon Britain

(*after E. T. Leeds*)

hungry land, too easily drying out after rain, which would be the farmers'
despair. They sought, as successive generations of farmers have sought,
the lands where clearance and ploughing would reveal good deep loams.
Such loams have sufficient clay to retain moisture and manure, but are
sufficiently sandy or silty to work easily and not to become sludge when
wet or hard clods when dry. They found what they wanted in the
brickearths overlying the valley gravels in terraces free from flooding
along the Thames and many other rivers : they found also what they
wanted in some of the chalky boulder clays of East Anglia ; they found

belts of fine mixed soils along the northern and southern flanks of the North Downs, and along the foot of the South Downs and the Chilterns.

The evidence of settlement afforded by place names confirms that of burial places. The early Anglo-Saxon kingdoms were focussed where such soils encouraged cultivation : the great clay tracts formed "negative" areas of no man's land between the settled tracts. Nearly everywhere the actual traces of the early settlements have disappeared. A few villages marked by hut-floors have been detected : the original hut walls were probably of mud and straw or at best of the 'cob' (stones, mud and straw) still used in the west country. One of the oldest of the English place name suffixes is *inga,* usually contracted to *ing,* meaning broadly a settlement, and where this is combined with a personal name of essentially continental origin it is clear evidence of early settlement—pre-seventh century. Other early suffixes are 'ham' (home) and 'ton' (enclosure, later village or town). Using the distribution of place names with -ing and -ingham and the Anglo-Saxon cemeteries we get evidence of the various early kingdoms.

The Kingdom of the South Saxons (Sus-sex) belongs to the later part of the 5th century and was confined on the north by the great Wealden clay forests.

The Kingdom of Kent, north-east of the same forest, was also apparently established by the end of the 5th century—by the Jutes. Many features differentiate the Kingdom of Kent from the rest of the country. The prosperous Kentish settler held a sulung (not a hide) as the area which could be ploughed by a team of eight oxen and it was divided into four yokes (i.e. each of two oxen). Primogeniture, which was the rule elsewhere, was replaced in Kent by gavelkind, a division of property among all first heirs. The Kentish village was not a compact unit : at a distance from the centre were forest clearings or *denns* for pigs and fertile marsh pasturage for sheep, as on Romney Marsh.

The Kingdom of the East Saxons (Es-sex) probably dates from 520 or 530 A.D. It was bounded on the north and north-west by the forested London Clay lowlands which formed an effective frontier from the East Anglian kingdom.

The East Anglian Kingdom (afterwards divided into the area of the South Folk (Suffolk) and of the North Folk (Norfolk) seems to date also from the early 6th century.

The origin of the West Sussex Kingdom (Wes-sex) is more obscure It may have had a focus originally in the upper Thames Valley, perhaps by settlers penetrating via the rivers of the Wash, but a bishopric was

FIG. 12

The Anglo-Saxon Kingdoms about the 8th century showing their relationship
with the modern counties

established later at Dorchester. It has been postulated that the Saxons
intermarried with and absorbed the earlier British elements. The Kingdom has survived in a sense that Wessex is an understood *nom de pays,*
though divided into the later shares or shires of Dorset, Hampshire and
Wiltshire.

Thus the early Anglian and Saxon Kingdoms extended inland only
as far as the Oolitic scarp—approximating in many respects with the old
Roman line of the Fosse Way from Lincoln to Cirencester. The heavy
soils of the Midlands were not only more remote but less attractive to the
early settlers. In due course however the middle Anglian Kingdom of
Mercia was established, in later years to be divided into shares or shires
—the now well-known Midland Shires. Northwards the Kingdoms of
Lindsey, Deira and Bernicia indicate the extent of the Anglo-Saxon
conquest and settlement of Lowland Britain.

We may return now to the general organisation of the Anglo-
Saxon settlement which gave us that unit so fundamental, so persistent
as to remain to the present day : the parish.

If the first vital need of the settlement was a water supply, we should
expect to find the village nucleus on one bank of a major stream,
athwart a minor stream, or close by a constant spring. Consequently
near at hand would be the common arable fields, sited on the best soil
available, preferably a loam terrace. Almost until the early 18th century saw the development of special fodder crops for animals, especially
root crops for winter feed, the farm unit or the agricultural village considered as a whole had to maintain a careful and proper balance
between arable, pasture and woodland corresponding with three
primary human needs : vegetable food for bread and drink, animal
food, and timber. Throughout English agricultural history there has
been competition between these needs. What is so frequently forgotten
is the very large acreage of pasture which was needed. Where an acre
of improved pasture at the present day, with all our knowledge of
nutritive strains of grass and clover, can support 0.4* stock unit, it
requires ten times as much rough grazing. Thus a Saxon eight-ox
team would have needed up to 200 acres of the rough grazing then
generally available. In reality the crucial point was winter feed—
hence the vital importance of hay from waterside meadows. Tansley
has drawn attention to the role played by cattle in clearing the land.
Turned out to graze in woodland the cattle eat some tree seedlings and
trample others, thus preventing natural regeneration so that rough

* See the calculations in Stamp (1954).

grazing with scattered trees and perhaps some thorny bushes replaces the woodland. Pigs turned out to eat acorns and beech mast were even more effective. The reverse effect is seen at the present day, when an ungrazed meadow or commonland where animals no longer graze passes quickly to scrub. Some even doubt therefore whether the Anglo-Saxons deliberately cleared much of the woodland : their animals did it for them. It is unfortunate that the misleading word "waste" has come to be used for grazing land absolutely vital to the life of the community. At first, of course, the poorer forested land on the valley sides was no one's interest: but as the village settlement grew a demarcation of boundaries became necessary and the delimitation of parishes resulted.

Such settlements were essentially self-sufficient. The village subsisted on the produce of the village tract. It is small wonder that the settlers turned their backs on the Roman roads and took small account of the cities those urban-minded people had left. More important was a place where villagers from a number of settlements could meet : a crossing place of natural routeways by land and water such as a good fording place. Naturally the villagers had need of farm roads, of trackways leading from the village to the common fields, the common grazing and the woodland beyond. Almost by accident the trackways of one village joined up with those of another and a through way from village to village was established. In this way the network of narrow winding roads linking the villages grew up in strange contrast to the straight lines of the geometrically minded Roman.

Just as in Indian villages today, or of a brief generation ago, each settlement had its essential craftsmen : the wheelwright, the carpenter, the smith. The few wants that had to be met from outside led to the growth of the market town within walking distance. With an occasional calf or some lambs, or a little surplus grain and some wool for sale, the market town took on other functions in the life of the countryside. In some cases it would seem that a market or fair was deliberately established on poor land bordering a number of villages which none of them in particular wanted. The site of Birmingham is often quoted as an example, though it is not a good one.

Thus, through the Dark Ages the humanized scenery of Lowland Britain took shape. As the Anglo-Saxons spread over what are now the Midlands the land was more homogeneous : the parishes are roughly circular with the village towards the centre (Fig. 15). On the whole the soils here are heavier, there being large stretches of Boulder Clay as well as of Triassic Marls or Liassic Clays. Any hollow in a field under such conditions will hold water and there are many surface streams. The water supply did not present the problem to early settlers here as it did elsewhere and farms could be located accordingly. To this day scattered farmsteads are the rule : many parishes have no single focus but instead a number of hamlets of roughly equal importance. On the other hand the settlers faced other difficulties. One was the arduous task of clearing heavy oakwood, another was the difficulty of deep mire on the track-ways in winter or wet weather. Even greater was the management of the clay soils. At a later stage these claylands became known as "three-horse" or "four-horse" lands according to the number of horses required to pull the plough compared with the one or two on lighter lands. Land drainage was a problem too and gradually the fields were ploughed into broad ridges with hollows between which served to drain off surplus water. In later times, especially in the latter half of the 19th century, many of these old ploughlands went down to permanent grass but the fields bear the marks of their Anglo-Saxon managers. The broad ridges give the whole land the appearance of having been permanently waved. (see Plates 2 and III). During the plough-up campaign of the Second World War much of this land was reploughed: modern large scale machinery ignored the old 'waves' with disastrous effects on drainage, and many a farmer (still more the County War Agricultural Executive Committee) had to re-learn Anglo-Saxon wisdom.

Rather naturally this settlement of the heavy lands of the Midlands came later than that of eastern England : over large parts Celtic place names suggest that an appreciable Celtic population lingered or that Celt and Anglo-Saxon worked side by side, or where the two groups actually fused. The period of expansion westwards was from the 7th century onwards : Wessex expanded westwards beyond its earlier limit on the western edge of Salisbury plain, whilst in due course the great central Kingdom of Mercia was established. Offa of Mercia devastated parts even of Wales, and set his boundary along Offa's Dyke, which is still to be traced. It was the later division for administrative purposes that gave us the Midland Shires.

Meanwhile two changes of major significance had taken place or were taking place in Anglo-Saxon Britain. One was the conversion of the people to Christianity and the establishment of churches in every village and town. The familiar story of Saint Gregory, last of the great officials of secular Rome and first of the mediæval popes, being fired by the sight of Anglian slaves for sale in Rome to undertake the conversion of England is too well known to need repetition. Unable to go himself, his plan materialized when Saint Augustine landed in England shortly before Easter, 597. Well received by King Ethelbert, the establishment of his headquarters at Canterbury dates from the same year, when the King gave him the old Roman church of St. Martin. What has become known as the Conversion belongs to the 7th century. The Anglo-Saxons built their churches as they built their homes, in non-durable materials —in cob and in wood. The very few Saxon churches that remain date from the immediate pre-Norman period, but there is little doubt that the parish structure was fully established when the Normans crystallized it for all time by church building in stone.

The other great change was the gradual evolution of the manor and the manorial system. The essential element in the English countryside, as over much of lowland Europe, was the village, and so it remained for a thousand years or more. It may be said that the village in England was the essential contribution of the Anglo-Saxon invader from the sixth century onwards. The whole organisation depended upon co-operation. With the cultivation of the open fields and the pasturing of animals on the common grazing there was no room for individualism : each cultivator, though he held scattered strips, must observe the same rules of husbandry, each the same dates of the agricultural calendar. The great advances in selective breeding of animals or experiments with new crops had to await the control made possible by separate enclosures or the private enterprise of the great landowner.

The settlement and development of Anglo-Saxon Britain had gone on steadily from about 450 A.D. for nearly four centuries before the new factor, the incursions of the Danes and the Norwegians, that is to say the Vikings, became a vital consideration.

The chief source of information on the Scandinavian invasion of Britain is the Anglo-Saxon Chronicle. Under the year 787 the Chronicler records that three ships arrived which were "the first ships of the Danishmen which sought the land of the English nation." It was not, however, until half a century later that the Scandinavian attacks on the coasts of the British Isles began in earnest. Danish troops wintered in Kent in 851 ; in 867 they seem to have conquered Yorkshire and in 876 the Chronicle records that the Danes parcelled out the land of Northumbria and began ploughing and cultivation—that is a permanent settlement. In 877 part of Mercia was seized, in 880 a part of the army settled in East Anglia and divided the land.

The story of King Alfred's resistance to the Danes, his failure and later his success is well known, at least in fable. The treaty which he concluded with Guthrum, King of East Anglia, in 886 (or it may have been a few years earlier) allowed that Danish territory (the Danelaw) was to be recognized as extending southwards to the lower Thames westwards to the Lea, then from the source of the Lea to Bedford, then from the Great Ouse to Watling Street—probably the greater part of northern England also (see Fig. 12).

From that period onwards there followed a gradual reconquest by the Anglo-Saxons and it is by no means clear what particular effect the Scandinavians had on the scenic evolution of Britain. Place names, notably those ending in -by (village or town or homestead), so widespread in the Midlands, or -thorpe (hamlet or daughter-settlement of an established village), and -thwaite (clearing) elsewhere indicate the wide extent of the settlement, but did the new invaders do other than take over the lands cleared and used by the Anglo-Saxons? In Scotland at least it would seem that much original clearing is to be attributed to the Scandinavian invaders and the origin perhaps of infield-outfield, or the run-rig system of agriculture. This is the system by which a small area near the farmstead or settlement is continuously cultivated and receives the manure from the animals (infield) whilst extensive shifting cultivation (five years in crops, five years fallow) is practised on land farther away (outfield). Both the infield and outfield were unenclosed in the modern sense but the Scottish unit was the farm rather than a village tract worked communally. A fifth or a quarter of the farm might be infield, the rest outfield, but in highland districts there was also a large area of moorland, rough grazing beyond the head dyke (wall or fence marking the limit of the "muir" and the farm proper). Such a holding might however have its open fields sub-divided amongst co-heirs into broad strips 18 to 50 feet wide separated by deep drainage furrows. This feature is known as runrig or rundale and suggests comparison with the Anglo-Saxon strip cropping. The whole Scottish system has been called the 'Celtic' system and certainly was destined to last until the early 19th century—indeed with comparatively little change. Was it Celtic or introduced by the Scandinavian invaders? Certainly in Wales there is little evidence of either the English open fields or the Scottish infield-outfield. Much more it would seem that in Wales the cultivation of small scattered fields was that which persisted. Gray, so clear on the Midland system of England, leaves us curiously confused in his determination to treat a Celtic field system as a whole. Plate II shows the arrangement of fields and the moorland edge in a typical Welsh valley.

Plate 3a. TEWKESBURY. A characteristic association of the Cedar of Lebanon (planted about 1790) with the Georgian-fronted Abbey house, now the residence of the Vicar of Tewkesbury   (*pp. 76-78, 185*)

b. AN 18TH-CENTURY MANSION IN MONMOUTHSHIRE. The grounds are embellished by a well-grown Sequoia (Wellingtonia) on the left and a Copper Beech (*right*) with clumps of rhododendron (*pp. 184, 209*)

(*L. Dudley Stamp*)

Plate 4a. THE COMMON DAISY (*Bellis perennis*). Here growing among long grass but flourishing best on lawns which are constantly mown and rolled. (*p. 100*)

(James Fisher)

b. A BUTTERCUP MEADOW. Meadows afford a wide extension of what would otherwise be a restricted habitat. The attractive flowers are not always those valued by the farmer and this could be called a neglected meadow (*p. 100*)

(New Naturalist Library)

Brief reference may now be made to the early stages in the development of the manor, so often regarded as a Norman innovation.

It is probable that the early stages of the evolution of the Norman manor can be ascribed to Danish influence. The Anglo-Saxons had come as settlers working the land on a communal basis. The Danes came as warring bands often under individual leadership. The leaders looked to the spoils of war for their enrichment, their followers expected rewards in grants of land and a note of individualism was introduced coupled with recognition of an overlord. Under the overlord or thegn* (himself recognizing the supremacy of the King) we find on the eve of the Norman Conquest freemen of different grades— geneats or socmen (sokemen), gebur and kotsetla (smallholders). Below these were the theow or slaves, often of alien, i.e. Celtic, origin. Before the arrival of the Normans the feudal system was already there in embryo.

As Professor H. C. Darby has emphasized†, "A half-servile community under a lord was not different in essence from a village of freemen, either in plan or in customary arrangement", and it was natural that the village of early settlement days should develop a leader to be replaced as the centuries passed by an hereditary headman who ensured the continuance of the system. Although the term "manor" came in with the Norman conquest, the great changes in the first decades of the conquest were in the persons of those who held the land from the King rather than in the system. We may presume that in later Anglo-Saxon times some village lands were still held by an association of freemen or sokemen, but in many the village was essentially in the possession or control of a thane. In Norman times, of course, the village became a manor and a lord of the manor was universal.

Whilst the two or three-field system was firmly established and lasted for centuries across the Midlands of England it either never existed or was eliminated in the east and south-east. It is highly probable that in all areas the energetic individual was free to clear a tract of woodland or plough a section of 'waste' for his own use, but a particular type of settlement developed in the remaining great woodland areas of the Weald and Essex. In the Weald place names ending in -*den* perpetuate the system, a denn being a forest clearing operated from an existing village settlement outside the forest zone. The suffix -*hurst*, indicating

*the spelling thegn is used by many historians instead of the conventional thane to avoid confusion with the Scottish use of thane.

†*An Historical Geography of England before A.D. 1800,* p. 191.

wood, is a perpetual reminder of the dominant position occupied by woodland in the same areas.

Other place names give a clue to dominant crops. These include *hwaete* (wheat) modernized as Whit-, *ryge* (rye), *claefer* (clover), *bean, pease* and *fleax* or *lin* (flax). A manuscript of the 10th century contains the first of all pictures of an English plough* which despite its heavy construction is shown being drawn by only two oxen. From other pictures in the same manuscript we learn something of the farmer's year— ploughing in January, breaking the clods by a spadelike and an adze-like iron-tipped implement in March, sowing broadcast in April. Under May a shepherd with horned rams of Exmoor Horn type and hornless ewes is shown ; June is occupied with haymaking, August with the harvest. Hay is shown being cut with a very modern looking scythe ; corn is harvested by a sickle just below the ears and small sheaves are shown being loaded into a well-built cart by modern-type pitchforks. Under September small-eared, stiff-bristled swine are shown feeding on the acorns and mast of the woods. Threshing and winnowing belong to December.

Perhaps the most important lesson of all to be gained is the picture of late pre-Norman England as "a country populous with a thriving peasantry, a land turned from a vast forest into a great sea of arable and pasture." These words are quoted from Trow-Smith, who goes further and refers to "pasture, so desperately short that it reduced the stint of stock to a low level . . . . . permanent hedges and fences may be postulated in many places and the classic picture of the hedgeless mediæval landscape will perhaps find itself in oblivion, along with grass baulk strip divisions and a universally rigid communal cultivation."

Certainly the evidence is steadily accumulating that Lowland Britain was, according to the agricultural knowledge and farming systems of the times, already densely populated, even overpopulated. Instead of there being any "waste" land, the natural woodland pasture was already so fully used that a strict system of stinting may already have been adopted (but see below, p. 94).

*Caedmon's Mss.

CHAPTER 4

DOMESDAY BRITAIN

NINETEEN YEARS after the victory at Battle near Hastings of William of Normandy in the memorable year 1066 the Anglo-Saxon Chronicle records tersely but vividly :

> Then at midwinter was the King at Gloucester with his wise men, and held there his court five days . . . . . Afterwards the King held a great council and very deep speech with his wise men about this land, how it is held, and with what men. He then sent his men over all England, into each shire, and caused them to find out how many hundred hides were within that shire, and what the King had himself of land and of cattle, and what rights he ought to have yearly from that shire. Also he caused them to write down how much land belonged to his archbishops, to his bishops, his abbots and his earls, and, though I tell it at length, what or how much each man that was settled on the land in England held in land and cattle, and how much it was worth. So very narrowly did he cause the survey to be made that there was not a single hide nor yardland, nor—it is shameful to relate that which he thought no shame to do—was there an ox, or a cow, or a swine left out that was not set down in his writing. And all the writings were brought to him afterwards.*

A contemporary account by Robert Losinga, Bishop of Hereford, confirms the record in the *Chronicle* and adds :

> Other investigations followed the first ; and men were sent into provinces which they did not know, and where they themselves were unknown, in order that they might be given the opportunity of checking the first description and if necessary of denouncing its authors as guilty to the King.

*J. Earle and C. Plummer. *Anglo-Saxon Chronicle, sub anno 1085,* 2 vols. Oxford, 1892-9, quoted by H. C. Darby, *The Domesday Geography of Eastern England,* Cambridge, 1952.

Transposed into modern terms one might almost be reading the story of the 1940-41 Farm Survey when the King's representative in the Ministry of Agriculture held a great council and very deep speech* with his wise men about this land . . . . . and then sent his men all over England into each shire and caused them to find out . . . . all about every one of the 300,000 farmers of the country . . . . . nor was there an ox, or a cow, or a swine left but that was not set down in his writing. And all these writings were brought to him afterwards . . . . . other investigations followed the first . . . . .

The parallel is more than just amusing. The original returns of the 1940 Farm Survey were and are confidential because they classify the farmers into A (good), B (average) and C (poor) which if published might be libellous and in any case the C farmers under the stressed conditions of World War II had to be denounced as guilty to the King's ministers. So the results of the Farm Survey were punched on to Hollerith cards : the Hollerith machine swallowed them up and in due course the King's clerks delivered up the useful if somewhat lifeless and soulless statistical summary.†

So too, some eight and a half centuries earlier King William's clerks, who would doubtless have rejoiced enormously in a Hollerith machine, had treated the original returns in much the same way. The stages by which they were rearranged and summarized and so fashioned into the Domesday Book will never be known. Many details, notably those about stock, were normally omitted. The rearrangement was intended, it would appear, to supply immediate information of the possessions of each owner. The book, says Robert Fitz-Neal, writing about 1179, "is called by the natives Domesdays—that is metaphorically speaking, the day of judgement." It was thus in fact a sort of Who's Who of the Landed Gentry, with full details of their estates.

As Professor Darby so aptly says, "In order to obtain a view of the English countryside in the eleventh century, the first task must be to undo the work of King William's clerks, and to restore the geographical basis of the survey by piecing together the severed fragments of each vill." Many partial attempts have been made to do this but it was left to Professor Darby to organize a band of researchers to undertake the gigantic task—though it is clear that both the inspiration and organization came from him and the bulk of the work was actually done by him.‡

*Actually in the air-raid shelter under 55 Whitehall.

†*Summary Report of the National Farm Survey* (1946).

‡Darby, *Op. cit.* (1952).

Plate V.—LONGFORD CASTLE, WILTSHIRE

Built by Sir Thomas Gorges, one of a famous family of the time, and his Swedish wife, 1573–1591. The original castle (left of picture) was modelled on that of Uranien-berg, Denmark. The garden is a good example of a formal garden of Elizabethan days, though remodelled in Victorian times

The Domesday Survey covered most of what in this book is called Lowland Britain except the north. The counties of Northumberland, Durham, Cumberland and Westmorland were not surveyed and only certain parts of Lancashire—which are sketchily considered as adjuncts to Cheshire and Yorkshire. Wales was also excluded but the counties of Cheshire, Salop, Hereford and Gloucester included lands which now lie within the borders of Wales. For the south-western counties of Cornwall Devon, Somerset, Dorset and Wiltshire there exists the *Liber Exoniensis* which seems to be a summary of the original returns and from which the relevant portions of the actual Domesday were compiled.

These details of the Domesday survey are given because they are needed to explain why it has proved so difficult up to the present, despite the immense amount of detail, to build up an adequate picture of England at the time.

Though the information is rarely complete and varies considerably from one area to another, Domesday gives material which may be classified as follows.

In the first place is the name and extent of the manor. The Norman-French word *manoir* in the first instance meant nothing more than a dwelling or residence. In parts of Domesday it is clearly used in this sense. For example in parts of Lincolnshire and East Anglia the "manors" are clearly very modest dwellings with several in a single parish or village. Even when somewhat larger, as in parts of the south, there was frequently more than one manor in a village. Where, however, the closely knit Anglo-Saxon village community had so come under Danish influence that it looked to a 'thane' as its head, then the thane's home was automatically the manor house and the extent of the "manor" coincided with that of the village tract or parish. Then there were cases where the lord owned or controlled land in more than one parish and the word manor became a convenient one to apply to the whole estate. In the words of Lady Stenton*, "The word manor had come before the end of this period to be almost a technical term. It meant an estate which was an economic unit, in which all the tenants were bound to the lord and his demesne farm, his free tenants paying him rent for their land and helping him at busy seasons : his unfree tenants doing weekly

English Society in the Early Middle Ages. Penguin, 1951, p.130.

Plate VI. FORMAL GARDENS AT STOKE EDITH, HEREFORDSHIRE
Laid out on a slope facing the house ; designed by Nesfield on the lines of London and Wise (1699). The edges are of box ; many compartments are filled with coloured earth (see page 77). (*Copyright Country Life*)

M.L.—E

labour service; and all of them regularly attending his court of justice, his hall moot, for the settlement of their quarrels and for the regulation of communal affairs."

At the time of the Domesday survey Danish influence over the old areas of the Danelaw was also marked by the numbers of free small-holders—descendants of Danish soldiers who had become farmers—many of whom held but a few acres but were free to sell their land or divide it into still smaller holdings amongst their children. Elsewhere large complex scattered estates had come into being called Sokes and although the peasants or "sokemen"(socmen) were free they and their lands were attached to the lord of the soke.

Domesday records the extent of the manor in terms of that variable unit of area already discussed, the hide.

In the second place the owner, or more strictly the holder, of the manor is recorded twice, in King Edward's time and in 1085-6. It is here that the hand of the conqueror is clearly seen: Anglo-Saxon and Danish names have in the great majority of cases given place to Nor-man. The record also includes the value of the manor in King Edward's time, at the time it was bestowed by William after the conquest and at the date of survey some twenty years later. The number of manors lying waste at the time of the survey, especially in the areas through which troops passed in the "harrying of the north" in 1069-70, is a measure of the devastation caused in rural Britain by the conquest.

In the third place there are details of the lord's tenants, their number, status and size of holdings or more particularly the number of ploughs. "Freemen" are often specified by name; much more numerous are the villeins, bordars and cottars who are together best described as unfree tenants of varying status though by no means slaves. Below them were the serfs.

In the fourth place are the details of land use, though these are expressed in such terms as to make calculations of actual areas very difficult. Wood-land is reckoned either as 'pannage' for so many swine or as so many leagues long and so many wide. Arable is usually stated as land for a given number of ploughs; meadowland and pasture more often in acres.

In the fifth place where appropriate are such details as those relating to mills, quarries and fisheries including, naturally, the annual value of each.

How far one can reconstruct maps of the period showing land use and, as a result, visualize the scenery of Domesday Britain has now been demonstrated by Professor Darby, first for East Anglia, later for other counties.

It should be noted that much of the value of the Domesday book lies in the picture it gives of late Anglo-Saxon Britain before the Conquest. The immediate effect of the Conquest was much devastation, seen to the full in the descriptions of 1086. In due course the land recovered so that the second picture was, fortunately, an ephemeral one.

Although the Normans came to Britain as aristocratic comquerors with little direct interest in farming, their influence on the development of the rural landscape was profound. In the first place, in order to establish their position as effective rulers, they stabilized and perfected the feudal system: at the top stood the barons who built those great castles which dominate to this day the strategic sites throughout the country.

In the second place, as briefly noted in the last chapter, they crystallized for ever the village pattern of England by rebuilding the Saxon wooden parish churches of the villages in stone. To some extent the difficulties of country planning today stem from this characteristic activity of the Normans and those who followed them. To destroy the village church in this land so rich in historic associations centring around it is unthinkable: yet to fill the pews and keep the pulpit occupied have become one of the great problems of our age. The Anglo-Saxon parish is a thousand years old: its boundaries were fixed by the Normans, and it does not necessarily fit the needs of today.

In the third place we owe much to the love of sport of the Norman barons and the 'forests' which they set aside for hunting. In our childhood many of us were brought up on the harrowing stories of William's men sacking the New Forest villages to make his hunting grounds. The picture has now been corrected: those settlers who had unwisely attempted the pioneer conquest of marginal land such as those hungry sandy soils of the New Forest, could scarcely have gone anywhere in southern England with less chance of farming success. Perhaps therefore their dispossession was for their own good: the same is true of the efforts of Scottish landowners in the last century to resettle the crofters from remote valleys where any development towards a modern standard of life was impossible. It was not until 1930 that our government compelled the 36 remaining St. Kildans to evacuate their remote island which many of them loved so well. Certainly the problem of marginal land is nothing new: the Normans appreciated far too well the value of good agricultural land to waste it even for the recreation of the most privileged. It has been left to the twentieth century to be indifferent to the God-given gift of good land.

CHAPTER 5

SCENIC EVOLUTION IN THE EARLY MIDDLE AGES

WILLIAM THE CONQUEROR was a great hunter. His passion for the chase he passed on to his son, William Rufus, and the pattern was set which the Kings and their nobles followed for many generations. It may indeed be said that from early Norman times onwards the land of England was divided into two parts. There were those settled parts to which the ordinary laws of the land applied, and there were those large areas created by arbitrary declaration of the King which were subject, in addition, to the harsh Forest Laws laid down equally arbitrarily by the King.

Just as in recent years the creation of National Parks in Britain has given rise to some extraordinary misconceptions as to what the term 'park' in this connection implies, so the term 'forest' has throughout the centuries been widely misinterpreted. A forest in this old sense is nothing more nor less than a tract usually of wild land, declared by the King to be subject to Forest Law. In studying the change in the scenery and use of land in Britain it is utterly wrong to consider the Royal Forests as uninterrupted woodland. Indeed, continuous woodland would not have constituted a satisfactory forest for whilst the wild animals of the chase need woodland for rest and for secrecy in breeding, the inclusion of "fruitful pastures" where food would be available to the deer throughout the year was absolutely vital. Though many of the Royal Forests of the Norman Kings doubtless carried far more timber than the areas do at the present day, the designation "forest" is not in itself any guarantee of a former woodland cover. It is doubtful, for example, whether Exmoor Forest or Dartmoor Forest were ever extensively timbered. Provided there were coverts of sufficient size and density to form breeding grounds, the greater part was probably open rough ground as it is today. The Royal Forests were not necessarily uninhabited. They could and did include villages and hamlets lying within forest clearings but of which the inhabitants were subject to Forest Law.

In days when the fastest means of travel was on horseback, the King sought to have hunting grounds conveniently located wherever he might be. Consequently forests were created throughout Lowland Britain wherever a tract of woodland and rough gave suitable conditions. That there was no forest in Kent, for example, may be taken as indicating that the country had already reached such a stage of settlement and cultivation that no suitable tract remained. One factor is, however, often forgotten. The Anglo-Saxon cultivators had sought out the loamy soils of the country—the loam-terrains—for cultivation. They were wise farmers : they avoided the light hungry sandy soils which accordingly were left in forest ; they likewise hesitated to attempt the task of clearing and managing the heaviest of clay soils (as in a large part of the Weald) which remained covered by dense, wet oak woodland. In summer the clay soils of the latter may be baked hard as bricks but for the greater part of the year the woodland trackways were deep in sticky mire; such land is not attractive to the huntsman so that there are or were considerable tracts of woodland, sparsely inhabited, which were *not* constituted legal forests. This is well illustrated by a comparative study of Figs 13 and 14.

It will be seen that many Royal Forests of the 13th century lie mainly on areas of light, poor soil (Category 9 of The Land Utilisation Survey) such as New Forest and Sherwood. Others, such as Exmoor and the Peak lie on mountain moorland (Category 8). Some are on medium quality light land (Category 5). Where the mediæval forests lay on rather better land they have in the course of later centuries given way to pressure of farming needs and have almost completely disappeared.

The animals protected by the forest law were the red deer, fallow deer, roe deer and wild boar. The last was virtually extinct by the end of the 13th century and a century later the roe deer ceased to enjoy protection because, it was claimed (in the reign of Edward III), it drove the other deer away. In Norman times wolves were a considerable nuisance in the forests, especially in the north, and those who killed them were sure of a reward. By the early part of the thirteenth century wolves had become scarce in the English forests though continuing long in Scotland. The King also encouraged the hunting of animals regarded as harmful to the deer by granting freely permissions to hunt the wolf, fox, cat, hare and even the badger and squirrel. Permission to establish a private forest was rare, but the King compensated his nobles by granting them "rights of warren" to hunt these lesser animals.

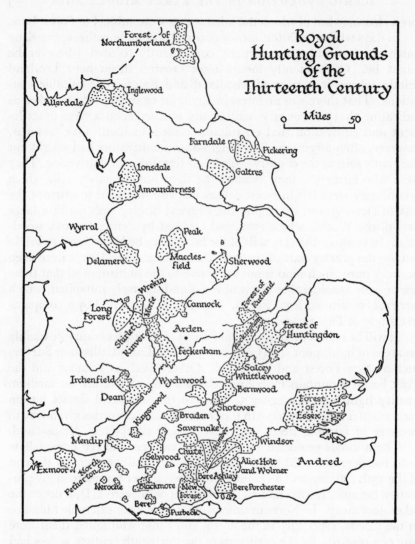

Royal Hunting Grounds of the Thirteenth Century

Miles 0 — 50

Forest of Northumberland

Allerdale

Inglewood

Farndale

Pickering

Lonsdale

Galtres

Amounderness

Wyrral

Peak

Delamere

Maccles-field

Sherwood

Wrekin

Forest of Rutland

Long Forest

Morfe

Cannock

Shirlet

Arden

Forest of Huntingdon

Kinver

Feckenham

Rockingham

Irchenfield

Wychwood

Salcey

Whittlewood

Dean

Kingswood

Bernwood

Shotover

Forest of Essex

Braden

Savernake

Mendip

Chute

Windsor

Exmoor

North Petherton

Selwood

Andred

Nerocle

Blackmore

Bere Ashley

New Forest

Alice Holt and Wolmer

Bere Porchester

Bere

Purbeck

FIG. 13
The Royal Hunting Lands of the 13th century (*after M. L. Bazely*)

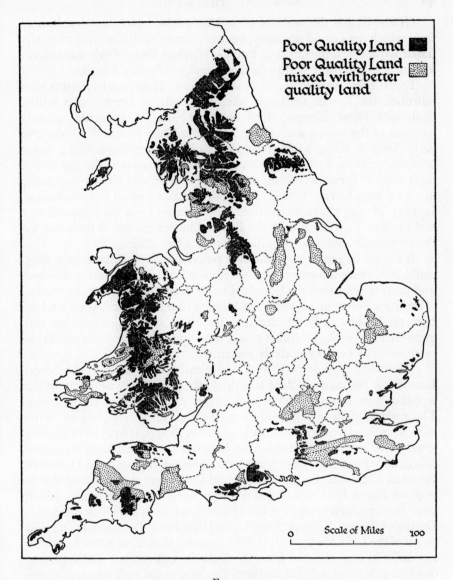

0   Scale of Miles   100

FIG. 14
The Poor Quality lands of England and Wales
Categories 7, 8, 9 and 10 of The Land Utilisation Survey (see page 246)

Gradually too the practice of making parks, which were enclosures from which deer could not stray, became common. They were often situated within or adjacent to a forest—Windsor Great Park was originally such an enclosure within the large area of Windsor Forest.

Forest Laws were strict, often savagely so. There were foresters who patrolled the forests, verderers who attended the forest courts which dealt with forest offences, wardens responsible for the actual administration of the forests and, at the head, the Chief Forester. Under the early Norman Kings a serf could lose his life for hunting a stag ; lesser offences, or those by freemen, involved penalties such as the loss of the right hand. Notwithstanding the penalties there was always poaching, always a smouldering hatred of the Forest Laws and all Forest officials. In 1215 Magna Carta established a reasonable basis for common law and justice. Two years later the Forest Charter did much the same for the forests : it abolished penalties involving loss of life or limb.

It should be noted that, however powerful the Norman Kings, they could not create a forest except where a suitable combination of wood and pasture remained uncleared. The oft repeated story of the creation of the New Forest involved the evacuation of certain hamlets which had been established in forest clearings but there is no evidence that this great stretch of light hungry soil had ever had any considerable population or been extensively cultivated.

The cultural scene in Britain today owes not a little to the Forest Law of the Normans. There is little doubt that otherwise many of the woodlands we so much enjoy would have been cleared of their timber. The greatest of those gifts from the past is the New Forest, but by virtue of its situation Epping Forest, tiny fragment though it may be of its once greater self, is scarcely less valuable. Windsor Great Park itself is another inheritance from early days : so too are Sherwood Forest and Delamere Forest in Cheshire. On the other hand the Forest of Arden and Rockingham Forest have disappeared except in name. The forests which were mainly rough open land have come down to us with little change— Dartmoor Forest, Exmoor Forest, the High Peak, Bowland Forest and much of the Lake District. It should be noted that these latter forests do not differ in any essential from the Deer Forests created under private ownership in Scotland and northern England in the 19th century, which are considered later (page 225).

"The English Kings by their forest law", says Lady Stenton, "were trying to preserve something that was bound to pass away. Civilized man lives by the plough, and not by hunting. The forests were bound to yield to the encroachment of the farmer." The encroachment took place in three main ways. In so far as the Royal Forests were concerned, the Kings from Richard I onwards saw the inevitability of the change but looked for a revenue from licensed "assarts" or tracts to be ploughed up. In the second place the Statute of Merton conceded in 1236 the right of the Lord of the Manor to enclose and plough up village waste—usually scrubland—provided he left enough common land to provide pasture for the village or commoners. In the third place woodland, scrub and rough pasture, in fact any no man's land not forming part of a Royal or private forest or chase, was a challenge to the individualist who felt himself unduly restricted by the communal system of the two-field or three-field village. Many of the isolated farms and hamlets (as opposed to nucleated villages) are present day reminders of this phase. The younger sons of farmers played a major part in this work of clearance. Since they were unlikely to inherit the family farm or the rights of their fathers in the village, they sought to establish themselves as squatters.

Historians have now come to recognize that undue attention has been given to the common fields of mediæval England—to the plough-lands of the village—and insufficient emphasis has been placed on pasture for cattle and sheep. Cattle were not bred primarily either for meat or milk : they were the work animals. The features to be prized were length of limb (especially in clay country) combined with heavy bone and development of muscle. The 'points' then to be considered were utterly different from those which the cattleman of today would list. The cattle had to be kept off the ploughlands until the date agreed by the villagers (Lammas) as marking the end of the harvest when they were turned into the aftermath. Waterside meadows were valued for their yield of hay for winter feed—almost the only winter feed available—and so it meant that for the greater part of the year the cattle had to graze elsewhere on the village lands. The unfortunate word "waste" has been and still is responsible for a great deal of misunderstanding. It would be better if we referred to "rough grazing" for it was on the unimproved but nevertheless vitally important village lands that the cattle found most of their food. In England we have become so familiar with our en-closed fields of permanent pasture that we forget that this form of land use is virtually unknown in many parts of the world. A Finnish farm may be found occupying a clearing in the pine-spruce-birch forests and despite regulations to the contrary it is in the margins of the forest that the cattle and horses of the farm will be found grazing. Or take a farm

in New England or the Eastern Townships of Quebec : it is often difficult to decide whether unploughed land should be called rough pasture,
scrub or open woodland, but it is there that cattle will be found grazing.
In mediæval England it is doubtful whether any extensive areas of the
lowlands remained unused by the farmer. If the open woodlands were
grazed by cattle, the denser oakwoods and beechwoods had long formed
pannage for pigs, where they fed on acorns or beechnuts and routed for
truffles and edible roots.

We still have to find living space for millions of sheep. The mediæval
farmer kept small short-woolled sheep on the hill pastures, and the
larger, heavier long-woolled sheep on the richer lowland pastures as well
as the ancestors of the down breeds. The mediæval sheep were bred and
valued primarily for their wool. Their skins provided parchment and
the farmer of the day understood both the value of their dung on ploughlands and their useful treading action in compacting light soils. Some
use was made of ewe's milk for cheese, especially on the marshes, but
sheep were not primarily bred for meat any more than were cows and
oxen. So much has been written of the mediæval wool trade and of the
wealth it brought to Britain that there is little need to repeat the story
here. The great churches erected by pious and grateful wool merchants
in the corn and sheep districts of East Anglia, the scarcely less opulent
churches of the Wold towns, the solid stone Cotswold villages are all
reminders of the wealth which came from wool and of some of the early
sheep-rearing areas. The great Yorkshire woollen industry of today is a
reminder of the shift to the hill lands of the Pennines ; the hosiery towns
of the Midlands, like Leicester, an indication that sheep rearing was
widespread there also. Indeed it was in the Midlands that sheep runs
replaced ploughlands. Estimates have been made of the numbers of
sheep. In the years between the two world wars England alone had
about 25,000,000 sheep. Bearing in mind the smaller animal and the
smaller fleece the numbers in the great days of the wool trade cannot
have been far below 8 or 10 million. Considering this number in relation
to unimproved character of the sheep grazings of the time, we reach the
inevitable conclusion that immense areas of the country must have been
given over to their needs. The better rough grazings of the Pennines and
Scottish Highlands today support about 1 sheep per acre ; on the
poorer each sheep requires ten to fifteen acres ; the natural fescue pastures of the Cotswolds and chalklands carry two or three. Although
sheep are able to survive the winter better than cattle, winter feed of
roots did not exist in mediæval times and we may hazard a guess that

the ten million sheep in the 14th and 15th century would use all the sustenance provided by ten or fifteen million acres. There is not much room for "waste" land. The change has been rather one of intensity of use (see below, Chapter 9).

Throughout the period of the early middle ages there is a contrast to be observed. It is something of a paradox that the heyday of feudal England should be the period when the essential keystone in the social structure was the almost self-contained village. In the later days of Saxon England the land had been divided for purposes of local government into shires. Each had a county or shire court to which all men of substance were bound to go and the "court" was the living repository of customary law. Their mouthpiece was the sheriff. Lesser disputes were settled at the lower level of the similarly constituted courts of the hundreds of wapentakes. At these each village was represented by the thane or leading men of the village. With the advent of William the Conqueror, all land was reorganized as becoming the King's property: he rewarded his followers by grants of land and titles which usually associated them with a county or shire. He encouraged them to build castles and to secure their position in an alien land where the King's strength depended upon the strength which his barons could build up. To the court and the baronial castles came the young men to serve their apprenticeship before becoming knights and returning to the land as Lords of the Manor. By the crystallization of the manorial system they in turn had a firm hold over both the land and the country folk who farmed it. So the unit in the whole system was the village, especially so when village and manor coincided.

The village was not only a closely knit community working as a unit but it had to be virtually self-sufficient*. Communications were so bad —the roads were mere trackways, often impassable for mud—that "exports" from a village were limited to wool, skins and hides, surplus corn and a few other commodities, such as products of the craftsmen, which could be taken to local markets or periodic fairs when weather conditions favoured travel. "Imports" were likewise few and restricted to such essentials as iron for the blacksmith to convert to ox and horse shoes, sickles, scythes and ploughshares, pots and cooking utensils; salt; and perhaps wine, spices and textiles for the great house. Where

* Recent historical research tends to show that the mediæval village was not as isolated as previously believed and that the roads were actually worse at the beginning of modern times. Nevertheless it doubtless remains broadly true that the lands within a parish produced the bulk of the food consumed.

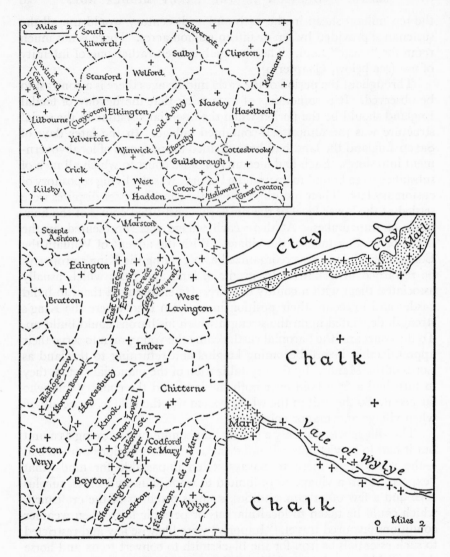

FIG. 15 (*above*) and FIG. 16 (*below*)

Maps of Parish Boundaries

In each case the cross marks the village church, usually at or near the village centre. In homogeneous lowland where surface water is everywhere available, the church or village is centrally placed and parishes are of much the same size. In country of contrasted types the church or village site was dictated by the need for a good water supply and the parishes are long narrow strips, taking in land of varied character.

a lord was in possession of several manors it was customary to move from one to the other to consume the produce, rather than attempt to bring produce to one centre from outlying manors. Similarly the court itself moved about the country. It follows that each village had to make the best use possible of its lands and in this sense the parish was the unit. From earliest times the good lands were known and used : markets and fairs took place on poor land of no great productive value. Just as the area implied by a "hide" was probably related to the productive capacity or inherent quality of the land so quite definitely was the size of the village tract or parish. Where the land is rich, parishes are small, for a small tract of good land could support a whole village with its church, parish priest and its whole range of rural craftsmen. The poorer the land the larger the area of the parish had to be. Where land of different types and qualities is found in close juxtaposition, the parish is essentially a heterogeneous unit. The village needed a good water supply, a tract of good meadow land, loamy soils for the arable fields, rough grazing and woodland for pasture. The shapes of such parishes as those bordering the Weald of Kent or stretching from the fertile valleys of Wiltshire up on to the chalk downlands—strips eight or ten times in length their width—are present-day reminders of the mediæval position. This is shown in Fig. 16.

The parishes which were thus aligned over lowland England at least in Anglo-Saxon or Danish times might not have persisted through the centuries until the present day had they not as noted in the last chapter been crystallized for all time by the building or rebuilding of the parish church in stone, a process begun by the Normans and continued in the early Middle Ages. In due course the mud huts or wooden homes of the villages were replaced by dwellings in local stone, or brick-and-timber where local stone was not available. One of the first dwellings to be given a substantial form was the manor house : many date from the 12th, 13th and 14th centuries, and were built to conform to the broad pattern, within a rectangle, of hall (one half) and sleeping chamber over buttery in the other half. A permanent substantial home for the village priest came later : at an early stage the village inn made its contribution to the village nucleus. With the increase of building in stone, bridges began to replace fords so that the road network, like the villages, was stabilized. Assarts and private enclosures began to fix too the pattern of the open land itself.

From this distant past we have inherited something approaching the 'normal' sized rural parish. On average quality land in the Midlands it tends to approximate to four square miles or 2,560 acres. The present day implications of this will be discussed later.

Feudal England and the manorial system lasted for some two and a half centuries after the Norman conquest of William the Conqueror to almost the middle of the 14th century. Changes were undoubtedly pending but the end of the manor was precipitated by the Black Death of 1348-9, when between a third and a half of the total population of the country died. The Black Death was a specially deadly outbreak of bubonic plague which is carried by the fleas of black rats (not by our now common brown rat). Naturally it struck most severely where the rats thrived amid the dirt of congested housing—in the slum or poor quarters of the towns and the labourers' hovels in the country. The upper classes in both town and country were accordingly less severely affected. It was the labourer who died.

Prior to that catastrophe the affairs of the village were controlled from the Manor Court which sat in the Hall of the Manor House. An annually elected jury of tenants administered the 'custom of the manor' and appointed—though it may be certain with the full approval of the Lord of the Manor—the various village officials. Surnames surviving in common use today remind us of village offices long since extinct. Thus the 'woodward' looked after the woods, the 'hayward' after fences, and the 'reeve' was a general assistant to the Lord's steward and who saw that allotted tasks were carried out. The 'shepherd' we still have ; 'cowherd', 'ox herd' and 'swineherd' are names which have become changed with time. All the villagers rendered certain services to the Manor, varying with the status of the villager, and in addition the Manor often had its own paid servants both indoor and on the demesne. The miller was an important man because the mill was the Lord's property and all corn had to be taken to the mill for grinding. The village craftsmen—blacksmiths, carpenters, wheelwrights, masons, brick and potmakers, basket weavers, and so on, completed the unit.

By the year 1300 certain changes had already taken place. It had become common for tenants to discharge their obligations to the Lord of the Manor by a money payment ; the Lord found it advantageous to let off his demesne as one or more separate farms for an annual rent. Actually there were fluctuations, money payments preponderating at some times, labour services at others. The demesne had frequently been enclosed—which was the lord's own affair and bothered no-one—but when extension of enclosure over the village common lands was undertaken opposition was natural.

After the Black Death, which wiped out whole familes, even whole manors, all over the countryside, land remained untilled, cottages and whole hamlets deserted. In some cases lands were left without any known owner or heir. Labour became very scarce; the labourer could not only bargain with his lord but even defied the village rules and migrated elsewhere. Efforts were made to enforce the old laws but the nationwide peasant revolt of 1381 if not immediately successful made certain the death of the old system. The Lord of the Manor became a landed proprietor, letting his land for rent. Shortage of labour, or the high cost of labour after the Black Death, had exactly the same effect as it had five centuries later—there was a swing over from arable farming and crop raising to pasture farming requiring less labour: in the 14th century it was to sheep farming. The increase in sheep farming continued for two centuries, 1350-1550, with the result that sheep enclosures though forbidden by Parliament, continued to increase; in years of poor harvest there was scarcely enough corn to feed the country. In some counties whole villages—probably already derelict—were swept away to make room for the sheep walks. The change had its compensations— wool brought money and with it imports from abroad and so, at least for those concerned with the trade, an increase in the standard of living.

A common mistake has long been to measure the status of a farmer in terms of arable acreage. This has probably never been true in any part of Britain, and was certainly no real measure in the late 14th and 15th centuries, at the height of the sheep-farming period.

At that time the 10,000,000 sheep postulated above in 1340 may have increased to 12,000,000 in England alone and that at a time when winter feed other than hay was unknown. There were great flocks owned by the monasteries, and by great landowners: probably as many, though less is known of them, by the newly arisen yeomen farmers. There were shortwoolled mountain breeds on the uplands of north and west, but the longwools which formed the basis of the export trade came from the now extinct Cotswold breed, from the Lincolns of the Lincolnshire heaths, the Romneys of the Kent marshes, and from the Midlands. The downland pastures, famed for sheep from time immemorial, had their breeds too. So highly was every scrap of grassland valued that even roadside verges and the unploughable turning points on common arrable fields were worth good money. In many parts of the Midlands only mounds today mark the sites of deserted villages overthrown by the pasture needs of sheep. Of course there is exaggeration in the plaints of contemporary writers, but there was a strong basis of fact when Sir Thomas More (1478-1535) said, " Sheep have become so great devourers and so wild that they eat up and swallow down the very men themselves. They consume, destroy and devour whole fields, houses and cities." In 1549 Protector Somerset introduced a poll tax for sheep, though it was never collected. Estimates of the numbers made at that time in connection with the tax, range from 8,000,000 to 11,000,000.

Such may have been the general picture of the centuries before the Elizabethan renaissance. Of course there were booms and depressions : there were efforts to increase both productivity and production by marling the land to convert hungry sands into fertile loams. There were efforts at co-operation and specialization amongst manors. As the church gathered strength many great monasteries practised what may be called high farming. The larger estates maintained their own roads and there was considerable movement of people and produce. In the countryside itself a new class had arisen—the wealthy peasant farmers and the earliest yeomen. There was considerable buying and selling of land, and it would seem to be clear from this that much land had been reclaimed and enclosed outside the common arable and grazing of the village where such existed. A two-roomed hut remained the lot of the poorer peasant, perhaps with a close or garden where vegetables were grown, bees were kept (for sugar was an expensive luxury and honey was used for making mead as well as for sweetening) and perhaps also poultry. The peasant had the right to use the common land for his pigs and for his sheep and cow if he were the lucky possessor of such. But he lived from hand to mouth : beasts were slaughtered and salted at Martinmas as there was little winter keep. January, February and March corresponded to the "hungry season" so well known amongst Africans in the marginal savanna lands today, and fasting in Lent was usually a necessity in any case. Till April and fresh vegetables put new life in the people, and perhaps life into the few remaining breeding stock, there was lack of fresh food, notably milk, butter and cheese. Scurvy was rife and the value of fruit, considered dangerous to health, was unknown even when it was available. Fish in demand for Friday and during Lent among the richer folk were costly to transport far from the coasts so that salt fish, carp and eels were commonly used.

At this time there was a close bond between town and country. Even the town merchant, dwelling for safety within the mediæval town walls, had there his garden, orchard, and even farm animals. Not unnaturally with increasing wealth he sought, as Englishmen have ever sought, a

Plate VII.—THE LANDSCAPED PARK OF STOWE, BUCKINGHAMSHIRE

Looking northwards across the centre lake to the mansion. Initiated in 1713 it was here that Capability Brown mastered the art of landscape gardening under the guidance of William Kent (*p. 78*)

The Time[s]

Plate VIIIa.—TRADITIONAL SHEEP PASTURES ON THE SOUTH DOWNS
A flock of Southdown sheep descending from the grazing on the Downs between Lewes
and Ditchling
b.—THE DOWNS RETURN TO CULTIVATION AFTER 2,000 YEARS
Part of the Ploughing campaign during the Second World War

Ford Motor Co. Ltd

place in the country, and so brought to the countryside a new element, but one destined to leave its mark in the succeeding centuries.

Once again the deduction is that during these centuries England at least, if not Wales and Scotland, was overpopulated: there were more people than the land could support at the technological level of the farming then known. It may even be that malnutrition was in large part responsible for the ravages of the Black Death. The stage was set for the economic changes associated with Tudor England, and especially with the Elizabethan age.

CHAPTER 6

THE FIRST ELIZABETHAN AGE*

THE DISCOVERY OF AMERICA by Christopher Columbus on behalf of Spain initiated a century of exploration, conquest and development without parallel in the world's history. Within the fifty years which followed, the Spanish *conquistadores* swept over the Americas from Mexico to the far south : with incredibly small forces accomplishing incredible feats of endurance they overthrew empires, establishing cities with churches, cathedrals and public buildings which are amongst the world's masterpieces of architecture even at the present day. The Portuguese, interested to a smaller extent in South America—in Brazil—were left free to pursue their older and equally crowded programme of exploration in the rest of the world. From the Guinea Coast, and rounding the Cape of Good Hope, they penetrated the farthermost corners of the Indian Ocean, and even into the Pacific. Within the first half of the 16th century they had established trading ports and forts as far afield as Cochin, Goa and Calicut in India, Galle in Ceylon, Malacca in Malaya, and Macao (1557) off the coast of China.

During these early years the British played a minor part, but with the accession to the throne of Queen Elizabeth in 1558 England's bid for the control of the seas was launched in earnest. With the defeat of the Spanish Armada in 1588 England became Mistress of the Seas, though only for a short period, and there were destined to follow three centuries of imperial expansion. Piratical plundering of Spanish galleons at sea or strongholds on land gave place to constructive exploration and the establishment of trading posts so that, even before the death of Elizabeth in 1603, the path was marked out. It was indeed during Elizabeth's reign that the British Empire was born. Not only was that phrase first used by John Dee but there was tangible expression given in the establishment of Raleigh's ill-fated colony of Virginia. The estab-

* I am greatly indebted to my colleague Professor F. J. Fisher for comments on this chapter.

lishment of the Levant Company, besides, marked an expansion of the great trading companies.

What effect, it may be asked, could these events possibly have on the development of British scenery?

There were some direct effects, admittedly perhaps of a minor character. The great demand for ships, both merchant and naval, then built largely of native oak, turned attention to the forests. The curved timbers needed in ship building had to be sought in the naturally curved limbs of the trees and even before felling the precise use of the timber in a given tree might well be determined. It may indeed be said that a characteristically English form of woodland management still widespread in the south today owes its firm hold to the needs of Elizabethan times. Only when an oak tree can grow with plenty of light and air does it spread its branches and produce the varied natural shapes so much needed by the old shipbuilders. So suitable young trees in a woodland were marked, others cut down, later to be allowed to produce numerous shoots from the stump : such is the original purpose of coppice-with standards*. If we no longer value the "standards", and have forgotten their original purpose, we appreciate the beauty of the noble old oaks in our favourite bluebell and anemone coppices.

In the words of William Harrison (1577) "Whereby the common saying is that 'no oak can grow so crooked but it falleth out to some use' and that necessary in the navy."

Shipbuilding stimulated a need for cannon, iron chains, anchors and many other articles of iron—the growth of the iron industry and the destruction of still more woodland in the demand for charcoal. Coal had not yet replaced charcoal except locally so that the salt works of Cheshire and Worcestershire are also recorded as using prodigious quantities of charcoal. Shortage of wood for fuel led to London's turn over to coal as the domestic fuel. Doubtless too the need for victualling ships stimulated certain types of farming near the main ports as well as the new naval stations which were springing up ; this was certainly the case in the Home Counties which supplied the expanding food market of London with its growing demand for bread, malt, meat, cheese, vegetables and poultry.

Then there were the indirect effects : the flow of wealth into Britain destined to continue for many years helped in the growth of wealthy

* An alternative explanation is the greater profitability of coppice wood which could be sold after 14-20 years' growth compared with timber taking 60-80 years to mature.

merchants who turned together with lawyers and politicians to the countryside, there to establish country homes worthy of their newly acquired wealth and glory. So the well known Elizabethan half-timbered homes of the clay districts, the stone mansions of the stone country, began to appear. Never before or since, it has been claimed, has the architect fitted his work so perfectly with the countryside. A new demand for timber from the woodlands followed ; a new develop-ment in ornamental parks and gardens. At the same time the yeomen farmers strengthened their position. Both they and the new merchant gentry wanted personal possession of land and hence enclosure. The luckless ones were the now landless peasantry.

The England of Queen Elizabeth has been called by Professor E. G. R. Taylor "Camden's England". William Camden was born about 1552 and his journeys through England were spread in the main over the last thirty years of Elizabeth's reign. Although his chief interest was in tracing the work of the Romans in Britain it is possible to build up from his great work *Britannia* at least an outline picture of Elizabethan England. The first edition (in Latin) of Camden's work appeared in 1586 ; the sixth Latin Edition in 1607. Translations by P. Holland were published in 1610 and 1637, and other translations (with addi-tions) later, notably by Bishop Edmund Gibson in 1695 and 1722.

Even more valuable in presenting us with a general picture of Britain in Shakespear's youth is William Harrison's account of Eliza-bethan England* which was written as the introduction to *Holinshed's Chronicles* (first published 1577, expanded in 1587). William Harrison, who lived from 1534 to 1593 was described by Withington, perhaps with some exaggeration, as "the only man who has ever given a detailed description of England and the English."

On the agricultural side there are two literary landmarks. One was the publication, first in 1523, of Anthony (or was it brother John ?) Fitzherbert's *Boke of Husbandry,* and the other the appearance in 1573 of Thomas Tusser's *Five Hundreth Points of Good Husbandry,* expanded from his *Hundreth Good Points* of 1557. They remained standard works for more than a century.

Fortunately the Elizabethan Era was also the era of the great map makers, Christopher Saxton (1542 or 1544—c. 1611) ; John Norden

---

*Readily available under this title in *The Scott Library,* No. 50, published by The Walter Scott Publishing Company, edited by Lothrop Withington, condensed from the edition of 1876, prepared for the New Shakespear Society by F. J. Furnivall (c. 1890 ?).

Plate IXa.—MECHANISATION ON HEAVY CLAY LANDS

Restoring to cultivation land unploughed in some cases since the Middle Ages. This picture shows members of the Women's Land Army at work during the Second World War

b.—HARVESTING OF OATS BY TRACTOR. Near Ranmore Common, Surrey

(1548-1625?) ; and John Speed (1552?—1629). With these three may be linked the name also of Philip Symonson who died in 1598, after publishing in 1596 a magnificent map of Kent. The county maps of Christopher Saxton which were collected together as an atlas published in 1579 have been made available to a wide public through the issue of a facsimile edition by the British Museum in 1936*. John Speed's maps of 1611 are similarly being reproduced in facsimile†.

At first sight it may seem that the amount of information regarding scenery and land use contained in the Elizabethan county maps is disappointingly small. The picturesque representation of mountains and hills, the quaint symbols for villages and manor houses as well as the rich embellishments scattered where space allows have rendered these maps highly prized in modern times for decorative purposes. Perhaps they are too highly prized, for atlases are commonly broken up into separate plates and sold framed. In this way they are, it is true, preserved and not destroyed, but no words are strong enough to condemn the vandals who cut up old maps for making into lampshades. Perhaps the words should be stronger still for those who encourage this destruction of priceless records by buying the articles so made.

When however the Elizabethan maps are read in conjunction with Camden and Harrison and especially when detailed samples are available, as they occasionally are, in old estate maps, it is possible to reconstruct far more effectively a picture of the surface of Britain at the time. Actually this has never yet been systematically and scientifically attempted : it is a task awaiting a patient researcher. It should for example be possible to delineate with considerable accuracy the parklands of the time and to settle the vexed questions regarding the area which they occupied and whether or not they seriously curtailed the productive use of farm land. Similarly it should be possible to get closer to the truth regarding the remaining extent of woodland and forest. Most writers of the period—as indeed before and after—are so clearly moved by their own views that their judgment is obviously subjective.

Such a subjective judgment is seen in Harrison's account of parks.

*An Atlas of England and Wales, 1574 to 1579, with introduction by E. Lynam.
†John Speed's England, Part 1 South-western England. Ed. by J. Arlott. London : Phoenix House, 1953.

Plate X. A FIELD OF WHEAT, HERTFORDSHIRE, AUGUST
In a typical English setting of fields divided by hedgerows with hedgerow trees and clumps of trees kept largely for amenity reasons. (*John Markham*)

He shall speak for himself : "In every shire of England there are great plenty of parks, whereof some here and there, to wit, well near to the number of two hundred, for her daily provision of that flesh, appertain to the prince (i.e. Queen Elizabeth), the rest to such of the nobility and gentlemen as have their lands and patrimonies lying in or near unto the same. I would gladly have set down the just number of these enclosures to be found in every county ; but, sith I cannot so do, it shall suffice to say that in Kent and Essex only are to the number of an hundred, and twenty in the bishopric of Durham, where great plenty of fallow deer is cherished and kept . . . . . our parks are generally enclosed with strong pales made of oak, of which kind of wood there is great store cherished in the woodland countries from time to time in each of them only for the maintenance of the said defence and safe keeping of the fallow deer from ranging about the country. Howbeit in times past divers have been fenced in with stone walls, especially in the times of the Romans, who first brought fallow deer into this land (as some conjecture), albeit those enclosures were overthrown again by the Saxons and Danes. Where no wood is they are also enclosed with piles of slate ; and thereto it is doubted of many whether our buck or doe are to be reckoned in wild or tame beasts or not . . . . . I find also the circuit of these enclosures in like manner certain oftentimes a walk of four or five miles, and some-times more or less. Whereby it is seen what store of ground is employed upon that vain commodity, which bringeth no manner of gain or profit to the owner." Harrison goes on to explain that venison is neither bought nor sold, and that it would be as derogatory for the nobility to contemplate selling their venison as it would "to degenerate from true nobility, and betake themselves to husbandry." "Here in times past", continues Harrison, "many large and wealthy occupiers were dwelling within the compass of some one park and thereby great plenty of corn and cattle seen and to be had among them, besides a more copious procreation of human issue, whereby the realm was always better furnished with able men to serve the prince in his affairs, now there is almost nothing kept but a sort of wild and savage beasts, cherished for pleasure and delight ; and yet some owners, still desirous to enlarge those grounds as either for the breed and feeding of cattle, do not let daily to take in more, not sparing the very commons whereupon many townships now and then do live, affirming that we have already too great store of people in England, and that youth by marrying too soon do nothing profit the country but fill it full of beggars to the hurt and utter undoing (they say) of the Commonwealth. Certes if it be not a

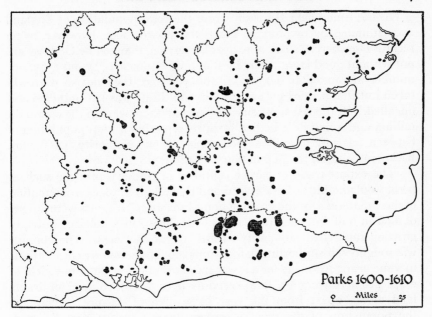

Parks 1600-1610

0 ___ Miles ___ 25

FIG. 17

Map of South-eastern England showing parks since A.D. 1600 as indicated on the
maps of Norden, Saxton and Speed

curse of the Lord to have our country converted in such sort, from the
furniture of mankind into the walks and shrouds of wild beasts, I know
not what to say. How many families also these great and small game
(i.e. deer and rabbits) have eaten up and are likely hereafter to devour,
some men may conjecture but many more lament, sith there is no hope
of restraint to be looked for in this behalf because the corruption is so
general." Harrison goes on in the same vein, warming to his task, till we
get cottages pulled down and not replaced, "cities and towns either
utterly decayed or more than a quarter or half diminished, though some
one be a little increased here and there, of towns pulled down for sheep-
walks". It was the "Normans, who added this calamity also to the
servitude of our nation, making men of the best sort furthermore to
become keepers of their game."

Harrison records that in addition to the private enclosed parks "we
have the frank chase, which taketh something of park and forest . . . . .
saving that . . . . . the chase (is) always open and nothing at all enclosed, as we see in Enfield and Malvern chases."

At the time when Harrison wrote the total population of England was estimated by two foreign observers at 3,500,000. However, as he records elsewhere, "as for able men for service, thanked be God, we are not without good store ; for, by the musters taken 1574 and 1575, our number amounted to 1,172,674, and yet were they not so narrowly taken but that a third part of this like multitude was left unbulled and uncalled." If the last supposition be correct there were in the country a million and a half able bodied men, representing a total population of between 4½ and 6 million (according to one authority 4,688,000). Rickman believed the population was 4,811,718 in 1600*.

The export trade of earlier centuries in primary produce such as corn, wool and hides, besides lead and tin had given place to a retention of these materials for the growing home industries. Imports were not yet of food in bulk but such luxuries as wine, linen, silks, tobacco (already growing rapidly in popularity), fruits, spices and sugar. The country was roughly self-supporting in foodstuffs though there were corn imports in some years. Allowing for a lower standard of living but a low yield of crops per acre we get a total certainly over 5,000,000 acres of arable land alone, apart from the very large area of pasture, to support the population of the day. A century later Gregory King in 1698 on an overestimated total acreage for England and Wales considered there were 9,000,000 acres of arable together with 12,000,000 acres of pasture and meadow at a time when various authorities estimated the population as between 5,200,000 and over 8,000,000. It is noteworthy that the general figure of rather more than an acre of arable per head of population is maintained. Incidentally King's estimate for parks and commons in 1698 totalled only 3,000,000 acres in England and Wales (including forests used for hunting). Even allowing for considerable dis-parking Harrison surely exaggerated.

Harrison devotes separate chapters to Gardens and Orchards, Food and Diet, Air and Soil and Commodities, Cattle Kept for Profit, Wild and Tame Fowls and Fish. In all we get a picture of an England not very different from that which we know in its more remote parts today. The balance between arable, pasture and woodland as supplying three of the primary needs of mankind was being maintained. The landed gentry—both the old families and the new merchant prince class—and the prosperous yeomen farmers between them farmed the land which they had tamed, improved and now owned. In this pattern of land use

*J. A. Rowse (1950) accepts the following figures : 4.5m in 1558 ; 5m in 1603 ; 5.5m in 1625.

there remained an anomaly handed down from the past. That anomaly was the unenclosed arable fields of the villagers and the common grazing on which they pastured their few poor animals. The arable fields had been laid out by the Anglo-Saxons who understood land and had selected the best. There is little doubt that over much of England where open fields existed the position had been reached that food-yields from poorer but enclosed land greatly exceeded that from the good lands which remained open. By Elizabethan times, the advantages of enclosure with control of grazing and stock and the reward of individual initiative instead of the pace of the whole being dictated by the pace of the slowest as in the open lands had been proved and were accepted by the progressive minority. Why then did enclosure take another two and a half centuries to complete? When enclosure came it undoubtedly brought hardship, great hardship, to those who became the landless agricultural workers but it laid the foundation for all real development in farming practice. Had the phrase been known at the time, open field village farming would have been stoutly defended as 'a way of life', 'traditional' and therefore praiseworthy and to be preserved. There seems an uncomfortable parallel with the present day advocacy of family farming as a way of life in face of demonstrable evidence of the greater efficiency of the large mechanized unit. But just as there are today highly efficient family farms so there were well-farmed open fields.

In due course the landless peasant was to provide the labour force when the industrial revolution got under weigh. In his thousands he migrated to the new industrial towns, there to live in the restricted environment of a back-to-back house and to forget for ever the ways of the countryside. Such was the initiation of the divorce between town and country. It is small wonder that the cry of the town was for so long cheap food and with little or no sympathy for those who produced it, where or how. In due course the entire power, the control over the national estate, has come into the hands of the town voter, outnumbering the rural by four or six to one at least. It is the descendant of the landless peasant, first dispossessed by enclosure, who now rules all.

CHAPTER 7

IMPROVERS ALL

WITH THE DEATH of Queen Elizabeth in 1603, the English throne was occupied by the first of the Stuarts—James I of England and VI of Scotland—thus securing the union under the crown of England, Wales and Scotland. The century which followed was to witness the pitting of strength of nobleman against yeoman, with civil war sweeping over the country and that republican interlude of eleven years so unnaturally strange to this country. To a considerable extent the life of the countryside went on steadily and seemingly unruffled by these violent upheavals in the political sphere. For the countryside it was, on the whole, a time of progressive rather than spectacular change. Although much has been written of an agrarian revolution it is difficult to draw any real parallel with the changes in the urban sphere which are called the Industrial Revolution. But one word crops up with increasing frequency in rural matters and that word is Improvement. It is symbolical of the simultaneous realization that agricultural practices had remained static far too long and that there was, indeed, room for improvement. The seventeenth century was interested mainly in arable farming: attention to stock, directly occasioned by the growing demand for meat in the new industrial towns, came in the eighteenth century.

Walter Blith published his *English Improver* in 1649 and expanded it in the *English Improver Improved* three years later. Fitzherbert and Tusser had held the field for a hundred years: Blith was destined to do so for the next hundred years. There is no correlation between the Improvers and either rank or political party. Blith's volumes were issued in the days of the Commonwealth, but it was Charles II only two years after his restoration in 1660 who became Patron of the new Royal Society (founded 1660) which immediately turned its attention to agricultural matters, including the setting up in its first year of a committee to investigate the possibilities of the potato. In 1663 Andrew Yarranton published his *Great Improvement of Lands by Clover**. The work of the improvers of stock in the next century I shall discuss in Chapter 11, but it is interesting to see how the idea of agricultural improvement remained uppermost in men's minds. It may be said to have culmitated in 1793 when the Government of the day set up the independent body entitled "The Board of Agriculture and Internal Improvement."

During this period of the Improvers home production of food normally remained above consumption and there was a net export, notably of wheat.

In 1601 Maxey estimated the arable land in England and Wales at 6¾ to 9 million acres. Billingsley quoted earlier estimates of 11,000,000 acres in 1688 and, if his estimates are near the truth, it remained at about this total for more than a century, Comber giving a total of 11,489,000 acres in 1808. During the period up to 1685 between a quarter and a third of the arable was under wheat yielding probably eight to ten bushels per acre. Imports were only allowed in times of real scarcity, exports only after exceptionally abundant harvests. After 1673 however an export bounty was granted whenever wheat prices dropped below 48 shillings a quarter of 8 bushels: the cultivation of wheat greatly expanded. In the 68 years 1697 to 1765 there were only four years in which, an import was needed, whereas exports averaged 1,648,000 bushels† of wheat a year‡. Then onwards the rising industrial population absorbed the home production. From 1793 onwards wheat imports became a regular feature of British economy. In 1807 Middleton estimated the net output per acre (deducting seed and wastage) at 16 bushels per acre, with an average per capita consumption of 8 bushels. This was probably a high estimate, and should be compared with 5.67 bushels in 1852 to 1896 and about the same later.

In so many directions the first Elizabethan age had seen the beginning of many changes destined fundamentally to alter and mould the British landscape, more particularly the English landscape. The advent of the Stuarts and the passage of the 17th century, despite its upheavals, saw the country on the threshold of great things.

In the first place was the development of the country mansion and its

---

*The volume published in this year was entitled "The Improvement improved by a Second Edition of the Great Improvement of Lands by Clover." Fussell was unable to find a "first edition."

†Winchester bushels, smaller than the Imperial bushel and approximating to the United States bushel.

‡Fluctuating harvests can be correlated with known climatic irregularities. There is no doubt that, as with animals, diseases played havoc with output in many years.

garden. From Saxon right to Elizabethan times the manor house had comprised essentially the lofty raftered hall on the one side, the buttery and sleeping chamber on the other, with much of the life and work centred in the courtyard. The Elizabethans introduced dining rooms and drawing rooms of single storey height, together with a range of bedrooms and chambers on the modern pattern. But from the point of view of the British countryside the outstanding change was the advent of the garden—the flower garden as distinct from the utilitarian fruit and vegetable garden. Through the reigns of James I and Charles I the love of gardening and of flowers which has become so characteristic of the English was introduced to them by Huguenot refugees from the Netherlands. The weavers who settled in Spitalfields started gardening societies: they introduced the tulip and a whole range of flowers now associated with the old English cottage garden—love-in-a-mist, honesty, nasturtiums and that stiff composite still called everlasting. The gardens of larger houses were essentially formal, set in rectangles and triangles with box and lavender edgings and hedges and walks embellished by statues and sundials forming part of a pattern which included also the house itself.

There were changes in the kitchen garden too, including the growth of a great wealth of herbs and a great interest in their use, but it was the addition of the potato to the garden vegetables which was destined to alter fundamentally the food habits of the British. We know all too little of the vegetables and fruits of the early kitchen gardens. The Romans were familiar with a wide variety of herbs and vegetables and may have introduced a number. It may safely be assumed that the cabbage, which is native to Britain and was extrolled by Pliny as the finest of vegetables, was cultivated. Since there are Anglo-Saxon names, derived from Latin, for onions, turnips, lettuce and parsley these probably date from the Roman occupation. Leeks were certainly popular with both Celts and Saxons—indeed the early gardener was known as a "leekward". The Saxons also had parsnips, garlic and mint—perhaps also beets and artichokes. By the 15th century radishes and carrots are commonly mentioned.

In mediæval gardens the vine, a Roman introduction, lingered on. Otherwise fruits seem to have been neglected. Cider apples and perry pears were widely cultivated but other fruit trees were restricted to a few noble gardens until about the 14th century, when Chaucer and Lydgate mention several varieties of eating apples and pears. Not till the late 12th century are there references to flowers being used to adorn gardens

Plate XIa.—Soay Sheep. Ailsa Craig (*p. 140*)

b.—Carting Hay with an Ox-cart, Fair Isle, September, 1937 (*p. 133*)

Plate XIIa.—Young Galloway Stores *Farmer & Stock-Breeder Photograph*

The shaggy coats of these cattle enable them to thrive under hard climatic conditions and even on grazing partly covered with snow. Holme Rose, Scotland (*p. 130*)

b.—Young Belted Galloways

Picturesque and characteristic animals. Upton-on-Severn (*p. 132*)

Farmer & Stock-Breeder Photograph

—roses were probably among the first—and not till the thirteenth century were wild strawberries and raspberries transplanted to gardens.

In the earliest known treatise on gardening by John Gardener (c. 1440 or 1450) over 80 vegetables and flowers are named as suitable for making sauces and stews. By the time of Henry VIII gardening had received a great stimulus : melons and "pompons" had become popular and a cauliflower—perhaps a rarity—is pictured in Gerard's *Herball* (1597).

By the time of Charles II cauliflowers were common, various peas and beans had been introduced, as well as Jerusalem artichokes, asparagus was esteemed and potatoes were common in gardens. In fact there are few vegetables we enjoy today which were not already in the gardens of three centuries ago. Indeed we have lost or forgotten some— such as the "skirret", a root sweeter than a parsnip—and as meat has improved make much less use of garlic and herbs. The main change has been an increased use of raw vegetables as salads (including the 19th century introduction of watercress) and development of celery and the still later introduction of the tomato. The latter, long known as the 'love apple' (or still more ambiguously as the 'wolf-peach') because of its supposedly aphrodisiac qualities, was brought from South America by the Spaniards but did not become common in England till late in the 19th century.

In his delightful book, *The English Garden*, Ralph Dutton distinguishes three periods in the development of the gardens. The first he refers to as the Search for Sustenance, 1066-1500, when the object of gardens was to increase the household food supply. Then followed the Age of Symmetry, 1500-1720, when the garden developed according to extremely formal patterns. Then followed The Return to Nature, 1720-1900, though murmurings against the current excessive formality were heard much earlier. Milton, a few years before the restoration of Charles II, painted a word picture in *Paradise Lost* of the Garden of Eden which clearly extols the natural garden. It was however William Kent (1685-1748) who introduced from his training as a painter the new concept into garden design that nature abhors a straight line. From 1727 onwards, when he worked with Horace Walpole in the development of the gardens for Chiswick Villa, he exercised an influence which paved the way for Lancelot Brown (1715-1783). Brown after working as a boy in kitchen gardens, eventually became controller of the Royal Gardens of Hampton Court and Windsor, a post which did not prevent him from designing or redesigning from 1750 onwards many of the great gardens of the country. When called in to advise he always found "great capabilities", hence his nickname of Capability Brown. Whilst he got right away from the straight lines of the formal garden he was no believer in wild nature—his work was to smooth nature, to dam streams, to grass their banks, to plant clumps of trees all designed to make a scene graceful, placid and extremely neat. Brown's work was followed by that of Humphrey Repton (1752-1818) who became a professional landscape gardener and, in 1785, was the first to use that term. In his revolt against the formal garden Brown went so far as to eliminate a garden in the ordinary sense entirely : the sweep of turf went right to the walls of the mansion itself. In other words, there was a park but no garden. On the whole however a distinction was maintained and persisted through the 19th century between a kitchen garden and a flower garden of more or less formal design, and a wild garden merging into the park where the area was sufficiently large. Naturally the Victorian age was a considerable return to formality—the monkey puzzle tree in the front lawns, beds with blue lobelia, yellow calceolaria and red geranium. This pattern still persists in many public parks. Some of the changes in garden design can be seen by comparing Plates V, VI and VII.

But to return to the 17th century and the Improvers, the problem of winter meat supply had not yet been solved and meat salted at the autumn slaughtering resulted in skin diseases, even scurvy, being prevalent. Quite apart from the scarcity of winter vegetables, the value of fruit and salads was not yet understood. The wealthier had dovecots and their ponds full of fish ; game birds shot or brought down by hawks were valued winter food.

In the fields the horse was gradually replacing the ox both for pulling the plough and for farm carts. We refer elsewhere to the influence of both Cromwell and the Royalist leaders in changing the character of the riding horses of the country. It was however too early for major attention to be given to farm livestock ; interest of the Improvers was centred on farm crops. What impresses one after a lapse of three centuries is not the pace of the agricultural revolution but the great lag in time between the introduction of a crop or a method shown by enthusiastic pioneers to be advantageous and its general adoption by the farming community. The potato long remained a garden curiosity : though introduced into Europe not long after the discovery of America and into England in Elizabeth's reign, it was not until 1699 that Worlidge hinted it might be grown in large quantities for swine and cattle. Arthur Young in his tours in the latter half of the 18th century thought it worthy of record when he found any considerable cultivation. Indeed its growth in popularity coincided with the rise of the urban population of the industrial revolution and the demand for cheap food.

Despite the obvious need for winter feed and better fodder crops the words and works of the pioneer improvers were not quickly heeded. Hay was still left till the seedheads had turned brown at the end of June : the value of clover had been proved but it was well on in the 18th century before its extended use became general. The seventeenth century improvers extolled the virtues of the deep-rooted "Holy-Hay" of the continent (Saint Foin) as well as Lucerne (La Lucerne) as providing fodder from thin light soils of sand and chalk lands. Lucerne persists for a life of 20 or 30 years but its acceptance has been slow. The natural pastures of the downs and marshes were valued but little was done to upgrade other grassland.

On poor arable land a two-course rotation with fallow in alternate years still persisted from Anglo-Saxon times ; on better land it was a three-course. There were many varieties of cereal and an increased use of manures, now including city and industrial waste. Marling was practised to counteract lightness of soil, spreading of chalk from local pits where available to counteract acidity. On peaty land, the surface was pared off and burnt, the ashes then being distributed.

Two great changes were still to come : mechanization or use of improved agricultural implements and the introduction of roots. The first is associated with the name of Jethro Tull, who published *The New Horse-Houghing Husbandry* in 1731. He invented the drill in 1701, though Worlidge had had the same idea half a century earlier. Tull's seed rows wide apart permitted cultivation and clearance of weeds between the rows and from every point of view this was an advance on the old broadcast sowing. He was a good practical farmer but a poor scientist, regarding all virtue as being contained in the soil. Fortunately this led him rightly to emphasize the importance of cultivation and the maintenance of a good tilth. Turnips had been known as a garden vegetable for centuries: not till Weston wrote in 1645 were they seriously advanced as a fodder crop, and it was left to Tull in the two decades before he died in 1741 to show the value of the root break in substitution for a wasteful fallow year. His teaching found a powerful disciple in Charles, Second Viscount Townshend, and from 1730 onwards Turnip Townshend revolutionized East Anglian farming by instituting the Norfolk four-course rotation of turnips folded with sheep ; barley or oats ; clover and rye grass ; wheat. Townshend only farmed for eight years : it was some years before his system was adopted but he had solved, with his fellow-disciples, the age-old problem of winter feed and so paved the way for the stock improvers and their work—Bakewell and the Collings brothers and the many who followed them (see Chapter 11).

In the meantime the face of Britain was changing. The country houses with their gardens and parks became more numerous. The parks were frequently made to serve as agricultural experimental stations for no longer was it derogatory to indulge in husbandry. England in particular had become a land of hedgerows and hedgerow trees giving place to stone walls in natural stone country. In many counties the predominant land use was a patchwork of small fields in the midst of which the unenclosed arable of the common fields and the common grazing remained as a survival from the past. The hedges were planted "quick", that is by using quick or live cuttings, and the "quick-set" hedge was usually of may or hawthorn (also known as whitethorn), sloe or blackthorn, crab apple, holly and elder. Ash and elm were often left to grow into hedgerow trees but other trees or shrubs, after growing for eight to ten years, were slashed and 'laid' to make a barrier remarkably impenetrable. It was often considered unlucky so to treat a holly tree : hence the many holly bushes which stand up above the hedge level. Naturally hedges were set along old lines—the margins of lanes, the borders of assarts or clearings, the limits of old cultivation strips—so that the field pattern of today enshrines centuries of rural history. As the fields were thus enclosed attention had to be paid to drainage and the ditch by the hedge became standard in many areas. So important is drainage that boundaries of properties are commonly defined not by the obvious hedge but by the less obvious ditch.

Apart from the enclosure of the common arable and common grazing which was still in many areas to take place, the eighteenth century saw what may be called the completion of the rural pattern by the people of the countryside themselves. What was to follow was in large measure the impact of town on country.

CHAPTER 8

THE INDUSTRIAL REVOLUTION AND THE COUNTRYSIDE

S O MUCH has been written of the Industrial Revolution in Britain that it would seem virtually impossible to add anything to the analyses which exist. We are here concerned however with a particular point of view : the effects which the Revolution had on the evolution of Britain's man-made scenery. The effects are not only numerous : they are sometimes direct and obvious, at other times curiously oblique and unexpected.

In broad general terms before the Industrial Revolution, the face of Britain had been slowly evolving towards a mature agricultural-rural pattern. In a new country the pioneer settler expends his youthful energy clearing the land of forest and scrub, or ploughing the prairie and planting the crops of his own choice. He does not yet know whether nature will work with him or against him. Nature will either make him or break him ; all over the world are those bleak or arid lands where nature has triumphed, pushed back the pioneer fringe, and again reigns supreme. On the North American continent the pioneer stage is only, at the most, a few generations old : farmlands cleared of boulders at huge cost in human labour can still be traced in the second growth woodland of New England; farm roads and deserted homesteads still show faintly from the air as one flies over water-hungry tracts of southern Nebraska. Elsewhere the inexhaustible fertility of loess soils in the Corn Belt is yielding ever greater returns in bushels per acre as intensive farming replaces the old demoded monoculture. It is perhaps hard to realize that this pioneer phase is fifteen or twenty centuries in the past where Britain is concerned and that the intricate land use pattern of today reflects with a considerable degree of accuracy land potential under existing forms of management and existing techniques.

This indeed was already the case to a large extent when the Industrial Revolution first made its influence felt. The Romans had behaved

much as white settlers in the tropics still behave today : they thought
they had what Americans call the 'know-how' and could afford to laugh
at the ignorant natives. Sometimes their decisions were wise : at other
times their mistakes show how little they understood the alien environment of Britain. Within seventeen miles of the centre of London, if one
is willing to brave the deep London clay mud or the bramble-thickets,
one can penetrate the heart of Ashtead Woods in Surrey, there to find
the remains of a Roman villa : perhaps the residence of a landowner-
farmer. He chose a ridge above the general level of the great oak-forest
which was wise, but he sited his residence on the most intractable type of
London Clay which was unwise. Doubtless it defied his efforts to clear
and cultivate and how he, from the sunny skies of Italy, must have
cursed the grey skies and the drizzle which gave him a foot of impassable
mud all round, and how in the advent of summer warmth and sun he
must have despaired of that same clay baked as hard as bricks and
cracked to a depth of a couple of feet. Clay land of this sort without
elaborate surface and sub-surface drainage and without chalking or
liming has remained the despair of farmers ever since : no doubt that
is why the Roman villa lies today in the depths of the woods.

The Anglo-Saxons and others who came in later centuries came as
settlers determined to till the land. Physical and climatic conditions
were not so very different in their homelands and they recognized the
good ploughlands—the brickearths over the gravel terraces—and the
waterside flood plains with a high water table which would make the
best grazings. They knew the value of a good water supply in siting
their homesteads : they realized how much land would be needed to
support a village on a communal basis of two or three open fields. By
the time the Normans came the more fertile parts of the country were
well known and Domesday shows how closely population density coincided with soil potential and how closely land values already reflected
the inherent qualities of the soil. A Domesday map of land values in
Devonshire could be used today, nine centuries later, to indicate
relative land values in that county with reasonable accuracy*. So the
position remained through Tudor and Stuart times : in general it was
the most fertile lands which supported the most population. When
Gregory King made his famous estimates of Britain's population in 1695
the cool, damp northern counties were the poorest in the land in both
wealth and population. The coal already worked on Tyneside was not
able to lift Northumberland and Durham above the poverty imposed

* See below, Figs. 43 to 45.

on a rural countryside by physical conditions, relief, climate, and soil, affecting farming. The same is true of Cumberland and much of Lancashire, Yorkshire and Pennine Derbyshire. By way of contrast light rainfall, sunny summers and wide ploughlands had brought wealth and population to Norfolk and a block of six agricultural counties, Hertfordshire, Bedfordshire, Buckinghamshire, Berkshire, Oxfordshire and Northamptonshire—all, except Berkshire, north of the Thames. Already, it is true, the influence of the London market, with more than half-a-million people, had advantageously affected the development of Middlesex, Surrey and Kent.

The great road system initiated by the Romans had long since been forgotten. Miry, foundrous trackways did duty for roads: it was this difficulty of movement which had forced every village or parish to become as nearly as possible a self-sufficient producing and consuming unit. Such produce in corn or livestock as was produced for sale had of necessity to be taken by pack animal or driven on foot to the nearest market and a journey of more than four miles each way, or at most six miles, was a serious undertaking. Longer journeys were reserved for special fairs, often at the county town. Fig. 18 shows each of the 17th century markets surrounded by a circle of 4 miles radius. Fig. 19 shows how few of the markets had survived in 1920-30. Each active market is shown surrounded by a circle of 6 miles radius.

So the rural pattern was almost complete, though enclosure was only partial. Where it had taken place there were the cottages of the farm workers often 'tied' to the farm; the farmstead, farm-buildings and farm fields with separating patches of woodland and perhaps common grazings. An irregular network of lanes linked farms and cottages with the village centre where church, rectory, inn, smithy and craftsmen's homes were often hard by the manor house or residence of the squire. Sometimes the village was a single nucleated whole, elsewhere split between hamlets—an arrangement often reflecting local differences in water supply. A river ford, a crossroads, had often afforded the original *raison d'etre* for the village settlement: with a superior degree of nodality permitting access from a number of villages the market town developed.

Such is still the essential pattern of agricultural-rural Britain. Enclosure has long since been complete, but village greens and commons remain: the crooked lanes are still there though they are all well metalled and tarred: the main roads still wind leisurely and have only been widened or straightened at great expense or much upset to the

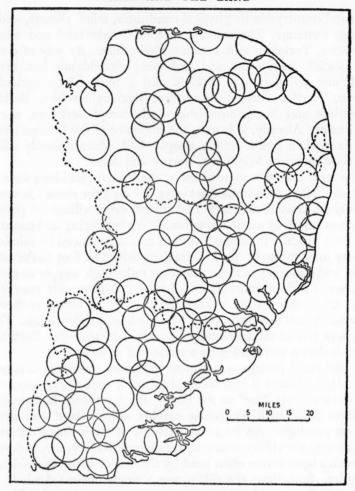

FIGS. 18 (*left*) and 19 (*facing*)
The Markets and Market Towns of East Anglia, past and recent (*after R. E. Dickinson*)

countryside. Increased access has concentrated life in the town rather than the village centre, but there is surprisingly little change in the number and layout of the farms and cottages.

Over this mature, age-old and seemingly stable land-use pattern there has spread another and alien one. It is the urban-industrial

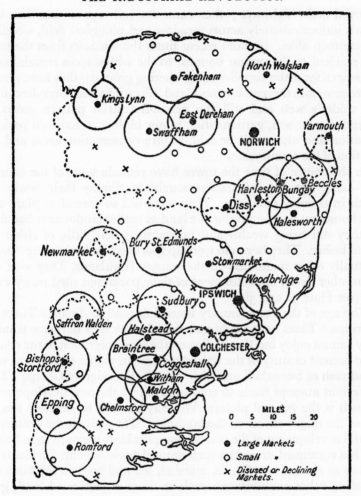

FIG. 19

pattern, initiated by the Industrial Revolution. Sometimes it combines with the older pattern, more often it absorbs, obliterates or completely devours it. Nine-tenths of the people of Britain are now part of that later pattern which has so completely altered the face of Britain. The spread of the industrial towns is an obvious phenomenon, but there is, in fact, scarcely any part of old rural Britain left untouched. At first canals cut their way across to link town and town and turnpike trusts improved main highways ; then came the railways cutting their knife wounds indiscriminately across meadow and ploughed field, woodland and common alike. In more recent times the tentacles from the towns have reached to the utmost bounds, to the widest spots remaining, in their search for water supplies and gathering grounds; they have planted their reservoirs in the remotest moorland. Electric pylons cover the country as a spider's web and still the urban-industrial pattern grows. It demands new towns, garden cities, motor highways, national parks for recreation and also, let it be said, nature conservation areas and preservation of the countryside.

In the course of time the towns have remade some of the countryside. Long ago the ironmasters sought to display their newly won wealth in country mansions. They were not so careful to plant their great houses on poor unproductive land as earlier landowners had done, but they did bring wealth and prosperity where little of either had existed before. They began to worship sport and build shooting "boxes" —actually other mansions—in the remote Highlands. They vied with one another in the embellishment of their parks and used many exotic trees (see Plate 3a and 3b).

The age of the great country house has passed. Income Taxes and Inheritance Taxes have made sure that the few may not have what the many cannot enjoy but there followed the mass migration to the country in the present century of the retired townsman with a consequent widespread rash of bungaloid growths. Not till the reign of George VI was any serious attempt made to control or direct the new development.

Such is the general picture. We may do well to examine some at least of the major effects on the landscape. Throughout the centuries of gradual development of the present complex land-use pattern in Britain, man has continued to restrict such natural environments for plants and animals as natural woodlands, marshes, fen and bogs, at the same time extending other environments such as those of cultivated land, grasslands, hedgerows, open woodland and wastelands. I shall discuss in a later chapter the way in which certain alien plants and animals have established themselves, but on the whole it is the native, especially amongst plants, which has seized the opportunity to spread itself more widely.

From Neolithic times onwards the cultivation of the land has given opportunities to weeds of cultivation and for centuries they were spread widely with seed-corn sown broadcast, nor were they eliminated by a fallow year. Many common field weeds were almost coextensive with ploughed land : others with land of certain types. Many will remember the blaze of red over acres of the lighter soils whether in Norfolk or Berkshire which denoted where the poppy had virtually taken possession of a cornfield. Fewer will recall the occasional spread of blue denoting abundance of the lovely cornflower—the *bleuri* of the French. The expansion of arable farming for bread production occasioned by the Industrial Revolution and sharpened by the needs of the Napoleonic wars, led to the expansion of the arable environment till the days of high farming in the eighteen-sixties and early 'seventies. Then a double change took place : the diminution of arable in face of the inflow of cheap imported grain, and the development of various methods of cleaning seed grain to eliminate the seeds of weeds. Today our corn-fields are cleaner—which means a loss of many once familiar as well as many unfamiliar cornfield weeds.

From very early times some natural grasslands were highly appreciated and carried a heavy stocking of sheep and cattle. This is true of the fescue pastures which afford the lovely springy turf of the downs ; it is true of the alluvial pastures of such tracts as Romney Marsh. When such pastures are subjected to controlled grazing and are adequately stocked, the animals after eating choice juicy grasses they like best, satisfy their hunger by eating the buttercups and daisies, the young thistles and the seedlings of shrubs and trees. The latter do not survive this treatment, many coarse weeds are so discouraged by being eaten that they sooner or later give up the unequal struggle. Many grasses, however, are simply stimulated to more vigorous vegetative growth—provided they are not overgrazed. Even autumn sown wheat which has grown too vigorously in the autumn can be grazed by sheep and will eventually produce a better corn crop as a result. For long little attention was given to the improvement of pastures which had been produced, in the main, by the grazing habits of animals. The older pasture farmers (as Pitt described in his *Agriculture of Leicestershire* in 1809) considered it a great achievement to have obtained in the long course of time a really good pasture and believed it to be sacrilege of the worst order to plough such a pasture. The idea of ley-farming, of 'taking the plough round the farm', and of sowing selected seeds mixtures—that is seeds of specially selected grasses—is a product mainly of the nineteenth century, and indeed, over much of England, has only been accepted since the Second World War.

CHAPTER 9

SOME FACTS, FIGURES AND FANCIES

IN THE PRECEDING eight chapters I have endeavoured to present a series of brief word pictures to show the successive stages in the development of what has been called the cultural landscape of Britain. In reviewing even so cursorily the results of man's handiwork two things seem to me to have emerged very clearly. In the first place the overriding and continuing influence of such physical factors as accessibility, the relief and elevation of the land, drainage, soil and climate became clear very early: the broad pattern of land use familiar today is already to be discerned in Roman times, was clearly established in Anglo-Saxon times and has scarcely changed over very large areas in the last two or three centuries. The use made of land of different types has changed in intensity rather than in kind. The lesson for the present day, in land-use planning which will be discussed later, is for us to respect the strength of the physical factors.

In the second place, at least from Norman times onwards and I suspect from earlier times, England certainly—the data for Wales and Scotland are more meagre—has been a densely populated country with a constant pressure of population on land resources. Perhaps it is my familiarity with some of the underdeveloped countries of the world today—in tropical Africa and south-east Asia—which has led me to see in mediæval Britain many of the same phenomena associated with land shortage, given the type of farming and level of technological development current at the time.

Necessity is the mother of invention : when population pressure increased to danger point, agriculture moved and technological development brought an increase in the home food supply. The two curves— of population increase and food output—rarely coincide : for centuries imports and exports took care of discrepancies but the mediæval export of wool (even of corn) must not necessarily be taken to indicate an overall surplus of primary produce. When the wool trade was at its height and wealth was flowing into the country there were many empty stomachs in the countryside.

For a time the agrarian revolution and resulting improved output of bread and meat kept pace with the industrial revolution and the food demands of the new towns. When the home population outstripped the ability of the homeland to supply the necessary food it no longer mattered. The later nineteenth century brought that automatic inflow of cheap food and raw materials from the new lands which drove the home farmer and countryman into a position of insignificance. The five or six per cent. of the people left on the land became at the worst country bumpkins who knew no better, or at the best Farmer Giles, a quaint survival from the past. Agriculture became a depressed industry, the countryside neglected and unkempt. This utterly false position was revealed in part by two world wars, but much of our thinking is still coloured by the position as it was in the Victorian era. Even today only some of our leaders are thinking afresh and trying to find a new balance and a new partnership between town and country, to consolidate the position of the nation in a changed world.

In the first place it is useful to have in mind some general figures of area—the extent of the land of Britain which is the homeland of its people. Since Roman times there have been accretions to the land area by the silting of creeks and reclamation of tidal marshes ; there have been losses by coastal erosion. As man has gained more of the mastery, no doubt gains have exceeded losses, but the changes have not been so great as to vitiate the use of present day areas as applicable to the past provided we remember that they included in the past large tracts of marsh, fen and bog then uninhabitable and since reclaimed.

Excluding inland water the official figures for area are as follows :

COUNTRY	ACRES	SQUARE MILES
England	32,033,000	50,051
Wales, including Monmouthshire ..	5,099,000	7,967
Scotland	19,069,000	29,795

The land area of England and Wales is thus roughly 37,000,000 acres ; of England alone 32,000,000. For purposes of comparisons with the past what may be considered the "effective" area of England was considerably less and coincided rather with what is called in this book "Lowland Britain". The limit of Roman Britain was Hadrian's Wall, which eliminates a very large part of Northumberland as it is at present. The Domesday Survey did not embrace Northumberland, Durham,

Cumberland, Westmorland, nor large parts of Lancashire, Cheshire and Yorkshire. In fact Anglo-Saxon England of the Domesday survey is roughly coextensive with Lowland England. Very approximately Domesday England thus defined may be regarded as covering 40,000 square miles, or 25,000,000 acres. This was the area which supported the bulk of England's population.

Turning now to population, I am greatly indebted to my colleague, Professor David V. Glass, for his kindness in preparing the following table showing estimated population for England and Wales since Domesday times. It is based in part on the figures assembled and analysed by J. C. Russell*, in part of Professor Glass's revision of Gregory King's estimate of 1695 and his own estimates for the eighteenth century. Figures for 1821 onwards are official census figures.

DATE			POPULATION IN MILLIONS		
			England	Wales	England and Wales
1086	(Domesday)	..	1 ·10	0 ·10	1 ·20
1340	(ca)	..	3 ·70	0 ·20	3 ·90
1377	..	..	2 ·23	0 ·13	2 ·36
1438	..	..	2 ·10	0 ·10?	2 ·20
1545	..	..	3 ·22	0 ·25	3 ·47
1695	..	..	—	—	5 ·2
1700	..	..	—	—	5 ·4
1750	..	..	—	—	6 ·2
1800	..	..	8 ·6	0 ·6	9 ·2
1821	..	..	—	—	12 ·0
1841	..	..	14 ·9	1 ·0	15 ·9
1861	..	..	18 ·8	1 ·3	20 ·1
1881	..	..	24 ·4	1 ·6	26 ·0
1901	..	..	31 ·4	2 ·1	32 ·5
1921	..	..	35 ·2	2 ·7	37 ·9
1931	..	..	37 ·4	2 ·6	40 ·0
1951	..	..	41 ·1	2 ·6	43 ·7
1961	..	..	43 ·4	2 ·7	46 ·1

Figures for Wales include Monmouthshire.

The population figures assume a special interest when we consider them in relation to the land areas involved. First we may take crude densities over England and Wales as a whole, with the following results:

*British Mediaeval Population, Albuquerque, N.M., 1948.

Date	Population (million)	Density Per Square Mile	Acres Per Head
Fourth century A.D. ..	0 ·5—1 ·5	9—26	71—25
1086 (*Domesday*) ..	1 ·20	21	30
1340 (*pre-Black Death*) ..	3 ·90	67	9
1377 (*post-Black Death*) ..	2 ·36	40	16
1430	2 ·20	38	17
1545 (*pre-Elizabethan*) ..	3 ·47	60	11
1695 (*Gregory King*) ..	5 ·2	90	7
1750	6 ·2	107	6
1800	9 ·2	158	4
1851 (*census*)	17 ·9	307	2 ·1
1901 (*census*)	32 ·5	558	1 ·2
1951 (*census*)	43 ·7	750	0 ·85
1961 (*census*)	46 ·1	790	0 ·81

In calculating densities and the land available in acres per head this table of course includes land of all sorts—the mountains, moorlands and swamps as well as the cultivable and cultivated lands. It is important to have some modern standards of comparison. If we take the world as a whole, excluding Antarctica as uninhabited and uninhabitable, at the present day the average population density is about 60 per square mile, or 11 acres per head. Already, then, by the 13th century population density in England and Wales exceeded that in the crowded world of today. Perhaps it is unfair, one may say, to take the world as a whole with its vast deserts and great mountains. Let us therefore take a country entirely in middle latitudes with none of the difficulties of dealing with frozen wastes or tropical lands, a country claiming that it is difficult to increase its population to any extent without a lowering of the standard of living. The United States had a population density of just over 50 at the census of 1960. England and Wales were already more crowded than this 700 years ago. Another interesting comparison is with countries having a climate like that of Britain and with a somewhat similar range of terrain. New Zealand had a population density of 23 at the census of 1961 and Tasmania between 13 and 14. It is highly probable that England and Wales were more densely peopled than this during the Roman occupation, and certainly in Norman times. Let us make one further comparison—with those countries of Africa

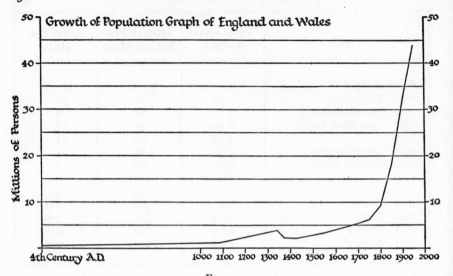

FIG. 20
Graph showing the rise in population of England and Wales, A.D. fourth
century to A.D. 1951. In 1961 total was 46·1 million.

which are at the present time presenting problems because of the
pressure of population on land. Here are some densities:—

Gambia	..	.. 75	Northern Rhodesia	9
Sierra Leone		.. 89	Kenya	.. 32
Ghana	..	.. 77	Uganda	.. 70

It will be noted that not one of these countries reaches the density of
our population in England and Wales as determined by Gregory King
at the end of the 17th century.

The general conclusion seems to me inevitable that if we make due
allowance for the levels of technological development at the periods in
question, Britain has experienced almost continuously from Roman
times onwards, certainly from Domesday onwards, a pressure of population on land to such a degree that the productive capacity of the land
was never more than a tiny fraction ahead of the food demands of the
population.

There are other ways of testing the general hypothesis that Britain in
the Middle Ages was suffering from severe pressure of population on
land resources. Taking the world as a whole at the present day there is
about 1.0—1.1 acre of cropped land per head of population. This

includes the extremely intensive cultivation of the Chinese (one fifth of mankind), the highly skilled farming of north-western Europe and the large-scale mechanized farming of the Americas. At the very high level of efficiency attained in present day Britain with wheat yields between 45 and 50 bushels per acre, the output of one acre supports roughly one person. Let us take as an example the period about 1340 before the Black Death. Wheat yields were 8 to 10 bushels per acre—let us say a quarter of the present day average—and other yields were in proportion.* Allowing for a much lower level of consumption the comparison suggests the produce of three acres would then have been needed to provide the food for one person. But with the three-field system one must add 50 per cent. for current fallows—4.5 acres per head†. If the population of England and Wales was then 3,900,000 one gets a total of 17,550,000 acres crops and fallow. But at that time there was an export of wool necessitating according to Trow-Smith a flock of 8,000,000 sheep without allowing any to supply the home market, perhaps another 2,000,000 sheep making 10,000,000. These animals were stocked on what today would be called rough pasture and a reasonable allowance would be 1·5 acres per sheep—15,000,000 acres in all. But the ploughing was done by cattle. Allowing each team of eight oxen could plough 180 acres in a season on the three-field system or 160 acres on the two-field, one gets about 800,000 oxen requiring 2,000,000 acres of improved pasture or 20,000,000 acres of rough grazing.‡

In those days there was, broadly speaking, no coal : wood was the

*The author of the *Anonymous Husbandry* probably compiled early in the 13th century gives some figures of yield, to be expected by the good farmer of the time. After suggesting that a reasonable sowing of seed in many places is 2.4 bushels per acre of wheat, rye, beans and peas or 4 bushels of barley or oats he suggests that yields should be :—

Barley	8-fold or, deducting seed,		28	bushels net per acre	
Rye	7-fold		14.4	,, ,, ,, ,,	
Rye-barley	6-fold	about	18	,, ,, ,, ,,	
Wheat-rye	6-fold	about	12	,, ,, ,, ,,	
Wheat	5-fold		9.6	,, ,, ,, ,,	
Oats	4-fold		12	,, ,, ,, ,,	

If one allows one-third of the arable land remaining in fallow the net overall production is reduced to two-thirds of these figures—and these are expectations from the good farmer, not actual realizations.

†Taking the Domesday population at 1.2 million and the same acreage per head this would give 5,400,000 acres under cultivation in 1086—very close to Seebohm's estimate of 5,000,000 acres.

‡In making these calculations I have considered 10 acres of rough grazing equivalent to one acre of modern improved grazing—the equivalent used by the Ministry of Agriculture for obtaining 'adjusted acreages' in the Farm Survey 1940-43—and then taken the stock carrying capacity of a modern improved pasture at 0.4 stock-unit per acre.

only fuel and charcoal for industrial use. May we hazard 5,000,000 acres of woodland ? What of the Royal Forests, chases and other land for sport ? At a guess these must have covered 5,000,000 acres. Allowing for working oxen 1,000,000 acres of improved and 10,000,000 acres of rough grazing and for cows and young stock a like total, if we total these land-use needs we get a total *requirement* of 64,550,000 acres. Fenland at that time had not been reclaimed and many mosses, now fertile land, were watery wastes. The total area of England and Wales was effectively less by about a million acres than the present total of 37,000,000 acres. So the hypothetical "requirement" of 64,550,000 acres had to be cut down to 36,000,000. Of course all these figures are the merest approximation and I am fully aware of the many considerations I have ignored. The corn export trade would have necessitated more arable land, the existence of good waterside meadows would have cut down the requirement of rough grazing. Light land would have been ploughed by a small number of oxen, and so on. But it is impossible to avoid the conclusion that there was intense pressure on land resources, and, *according to the types of farming used at the time* England and Wales were over-populated. There is an obvious need to make the qualification italicized in the last sentence : we make it at the present day in discussing pressure of population on land in tropical Africa in having to accept the current system of shifting cultivation or pastoral nomadism. The attempt to adjust stocking to land resources in the fourteenth or fifteenth century is clearly reflected in the use of "stints"—a limitation to the number of animals any individual could pasture on the common pasture, and described by Fitzherbert in 1523 as clearly of earlier origin.* In sum we may almost go so far as to say there was no "waste" in the country in the fourteenth century.

Throughout these calculations I have referred to England and Wales. Unfortunately the population figures available so far as I know for Scotland are too meagre for me to base similar calculations on them. In the case of England and Wales it might be better to separate off Wales and the mountainous parts of northern England, as suggested above. Over the remaining area of about 25,000,000 acres were concentrated, taking again the 1340 position, probably 3,500,000 out of the 3,900,000 total for England and Wales. Perhaps 14,000,000 acres of arable and current fallows would have sufficed, leaving only 11,000,000 acres for all other purposes.

*The Oxford Dictionary gives 1437 as the date when the word was first used in this sense. Compare the comments made above, p. 46.

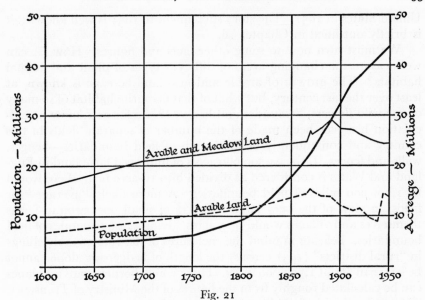

Fig. 21

The relationship between population and improved farm land in England
and Wales, 1600-1950. The trend continues.

One may make similar calculations for subsequent dates. Great
economies in land were made when improved pastures replaced rough
grazing (with sowing of rye grass and clovers after 1700), with enclosure
and the reduction in fallow or the introduction of rotations, with
increased yields and new crops. Production kept pace with needs: new
needs such as from the growing industrial towns of the 18th century
stimulated the farmers as we have already seen.

When did population finally overtake home production? At the
present day one acre of improved farm land in Britain is able to support
one person at our current standard of living and with our current level
of farming practice. We can feed 24,500,000 of our 46,000,000 people
(1961) which approximates to the post-war "target" of 55 per cent. of
consumption home produced. We have lost much land to the urban
growth: the 25,000,000 improved acres of 1850 at present level of
production would have supported 25,000,000 people when the population was actually under 20,000,000. We were in fact then able to live
comfortably and well on home production of food just at the time when
world wide markets were open for British manufactured goods. In
fact Britain's position in 1851 was closely comparable with that of the

United States in 1951. The story subsequent to 1851 is well known: it is briefly out-lined in Chapter 16.

We must turn now to some other facts and figures. How far can we measure the changing areas of different types of plant and animal habitats? The growth of arable and grassland acreage is known, at least over the last century, but what of that favourite habitat of so many of our wild flowers, "waysides and hedges"? So far as I know no calculation has ever been made of the number of separate fields in this country and consequently of the length of field boundaries—hedges, walls and fences. If the 24,500,000 acres of improved farm land in England and Wales is considered as divided into 10-acre fields there would be 1,200,000 miles of field boundaries. A more likely "average" for fields, in view of the very large numbers of small ones around farms themselves and villages would be 5 acres, giving 1,700,000 miles of field boundaries. Bearing in mind the enclosures of the 2,000,000 dwellings in "rural districts" (1931 census) the length of hedgerows alone cannot be less than about 1,500,000 miles. Those which border roads and lanes can be calculated roughly from the figures of the Ministry of Transport. The total mileage of public roads in Great Britain is 186,000 (159,000 miles in England and Wales, 27,000 in Scotland). Out of the total 8,250 miles are Trunk Roads, 19,500 are Class I, 18,000 are Class II or 'B' roads, and 49,000 are Class III. If we make an addition for private lanes and a deduction for town roads we get something of the order of a quarter of a million miles of hedgerows bordering roads—with all the familiar modern means of seed distribution.*

Whereas roads are open to the public, railway land is normally forbidden to the public. Railway embankments and especially cuttings have become almost natural reserves for wild life and since the track mileage of British Railways is approximately 20,000 miles the refuge is an extensive one.

In the great canal era of the late eighteenth and early nineteenth centuries most of the navigable waterways of Britain were interconnected, and gave scope to the wide distribution of the inhabitants of slow-moving or stagnant water. Today there are still over 2,000 miles of navigable waterways.

In a fascinating book published many years ago entitled *Man as a Geological Agent* the late Dr. R. L. Sherlock gave details of man's work

---

*A survey of field boundaries carried out by the Forestry Commission in 1954-7 found 954,000 miles in *rural* Britain of which 65 per cent. were hedges, 20 per cent. timber fencing, 15 per cent. other.

in mining and quarrying. Two aspects affect both the scenery of
Britain and the provision of habitats. One is the creation of much
"waste" land both in old quarries and more especially in quarry and
mine dumps which provided virgin areas of many different types for
plant colonization which would otherwise not exist. The other is the
creation first of conspicuous 'scars' on the natural landscape—one
thinks especially of chalk pits and quarries for road metal raised high
above the general land level so that gravity can be used in extraction—
which with later abandonment become grown over and again provide
primitive refuges otherwise rare.

As a prelude to the chapters which follow we may do well to remind
ourselves of the use which man today makes of the humanized landscape. So far as land use is concerned the following figures, in millions
of acres, give the broad position in 1960:

	England and Wales	Scotland	Great Britain
Total Area	37.1	19.1	56.2
Arable	13.7	3.1	16.8
Permanent Grass	10.8	1.2	12.0
Rough Grazing ..	5.0	12.5	17.5
Woodland ..	2.5	1.6	4.1
Other Land ..	5.1	0.7	5.8

In round figures there are 11,000,000 cattle in Britain, a third being
dairy cows in milk; over 28,000,000 sheep, the number having again
reached the old pre-war peak of some 28,000,000; about 5,000,000
pigs, though that is a number which fluctuates widely; and something
like 100,000,000 poultry.* What is important is the intensity of stocking of British farms and farmlands and the very high level of output per
acre. In output per acre Britain is beaten only, amongst the countries of
the western world, by Denmark, the Netherlands and Belgium. For
those interested in these aspects of British land use I would refer to my
book *Our Developing World* and to my larger work *The Land of Britain*,
3rd Edition, 1963.

*Sources of statistics for those interested are noted in the bibliography.

CHAPTER 10

CERES: SELECTION, NATURAL AND UNNATURAL, IN THE FIELDS*

NOTHING SURPRISES the observant American visitor to England who chances to land at Liverpool and journeys by rail through the Midlands to London more than the succession of small, hedged fields containing just grass. Coming from a country where, broadly speaking, a grass field, lush for a short early summer period only, brown and uninviting for most of the rest of the year, is an indication of land not good enough to plough and crop, he is at a loss to understand this apparent failure to make better use of land in a country so crowded, where home food production is so vital. It is difficult for the American, and indeed for many other visitors to our shores, to appreciate that the mild moist climate with its high humidity and low sunshine is first a deciduous woodland climate, but second, where that natural woodland has been cleared, is the climate that favours the growth of nutritious fodder grasses. It does not, except in the drier east, favour the ripening of cereals. Throughout practically the whole country grassland remains green throughout the year : no late hot summer scorches it brown, nor zero winters drain away its winter greenness. In the milder west where January temperatures average 42°F. or more, and grass growth goes on throughout the year, there is some grazing to be had by cattle and sheep even in midwinter. True, in the colder east and in Scotland stock may be brought in for the winter and stall-fed, but even there the fields remain green.

In the nineteen-thirties for every ploughed field in the lowlands there were two in permanent grass. As much of the ploughed land was sown to grasses (temporary grass) the townsman in many parts of the country formed the impression that the farms were a hundred per cent.

*For valued comments on this chapter I am greatly indebted to Mr. John Gilmour and Dr. Max Walters as well as to Mr. F. Perring and Mr. A. Lazenby of the Cambridge School of Agriculture.

grassland. In many parts of the Midlands he was right. Though the wartime ploughing campaign reversed the proportion—two fields ploughed to one of permanent grass—it is still the case on any count that grasses and associated herbs dominate over the bulk of Britain's farm lands. In the drier east of England temporary grasses make up one year in a three or four-course rotation ; in Scotland it is commonly three years of grass in a six-course ; in the wetter west of Britain ley farming involves three years in crops and some seven to ten years in grass. This is apart from the permanent grass of the alluvial marshes and riverside flood-plains where a high water-table or winter-flooding makes ploughing difficult or impossible, and apart from the claylands where grass in the last century has become almost traditional.

A distinction must be drawn between the upland or hill pastures and the lowland grassland in enclosed fields. The hill pastures—the rough grazings of official statistics—may be described as semi-natural : except in so far as they have been modified by grazing animals, especially sheep, they represent the natural vegetation of the country above the tree-line or where similar vegetation has replaced cleared woodland. At the highest levels lichens and moss with bilberry and some sheep fescue are found ; on mist-soaked ill-drained, high, but more level surfaces over peat, the cotton-grass (*Eriophorum*) predominates. On lower slopes are large areas of *Nardus* (Dry Grass Moors), a grass growing in tough masses unpalatable to sheep except for a short time in early summer before the excessive development of fibre. Similarly *Molinia* (flying bent or purple moor grass) which covers large areas of rather wetter land (Wet Grass Moors) and gives them a blue-green or purplish tinge grows so quickly in early summer as soon to become coarse and unpalatable whilst smothering the slower-growing better grasses and herbs. Both *Nardus* and *Molinia* mountain grasslands are improved from man's point of view by regular burning which gives other and more palatable grasses a chance to develop. The best or most readily improvable of the hill pastures are those where sheep's fescue (*Festuca ovina*) is an important constituent. This variable grass grows under a wide range of conditions : on wet mountain sides it is often the viviparous form which has the ability to produce young plants in place of flowers and seeds and so to spread where rainfall is heavy and sunshine scarce. Various species of *Agrostis* are associated with these hill pastures ; invasion by rushes (*Juncus* spp.) and sedges (*Carex* spp.) is indicative of poor drainage ; invasion by bracken (*Pteridium aquilinum*) an indication of better soil. The fescue mountain pastures are allied to the best of all natural

British pastures, those of the chalk downs and limestone hills where sheep's fescue (*Festuca ovina*) and red fescue (*F. rubra*) give a fine, matted, springy turf, and there is also a rich associated flora of herbs and other grasses. Most of the mountain pastures are on soil with an acid reaction, hence the ecologists' use of the term acidic grasslands : the downland pastures on the contrary are termed basic grasslands because the soil is basic in reaction.

By contrast the lowland enclosed pastures and meadows are not natural : they can scarcely be described as semi-natural, though the grasses and herbs of which they are composed are usually of native species. Originally the grasses and herbs must have had a restricted range in a countryside covered with woodland. As man cleared the woodland and his animals prevented tree regeneration by treading and grazing, so grasses and light-loving herbaceous plants, previously of restricted distribution, spread over the large areas of favourable habitat. The sequence of vegetation strips may be seen today on the margins of a broad farm-road : on the woodland margin or forming a hedge are low-growing shrubs or small trees such as hawthorn, blackthorn, wild rose and bramble : then tall vigorous grasses such as the taller species of *Festuca,* cocksfoot (*Dactylis glomerata*) and tussock grass (*Deschampsia caespitosa*) as well as *Agrostis*. Towards the roadway are the lower-growing grasses and those plants which seem to benefit from constant treading under the foot of man or even under the rolling action of motor tyres. The well rolled and constantly mown (but not weeded) lawn affords the extreme case with its abundant growth of the tiny daisy (*Bellis perennis*) and the Greater Plantain (*Plantago major*) with its hard-ribbed leaves pressed close to the ground.

If a grass field is neglected, the reverse change soon takes place. Hawthorn, blackthorn and brambles push out from the hedges, the finer meadow grasses are soon smothered, seedlings of forest trees soon appear, notably hazel and birch, so that within a very few years the meadow has become a thicket of scrub and coarse grass and well on its way back to forest. Yet this simple and inevitable change is little understood and always seems to surprise the townsman. A town council will purchase a group of fields or acquire the rights of a common as a public open space : the land which was so attractive when grazed by the farmer's cows and sheep soon becomes a dreary useless waste when they are removed.

Most of the ordinary permanent or long-ley grassland of lowland Britain is found growing on soils which do not depart widely from neutral in reaction, hence the term "neutral grassland", commonly used by ecologists. When heavy rainfall or prevalent mists tend to cause acidity the pasture deteriorates from the farmer's point of view unless the acidity is countered by dressings of lime in some form—as chalk, ground limestone, or slaked lime.

Although over a hundred species of grasses are recognized by botanists as native to Britain only about a dozen are really important from the agricultural point of view. Two grasses are so widespread and important that they have been used by Sir George Stapledon and Dr. William Davies as indices of pasture quality. These are Perennial rye-grass (*Lolium perenne*) taken as an index of the best or most carefully managed pastures, where it is commonly associated with wild white clover, and the common bent (*Agrostis tenuis*) which steadily increases in proportion in the less valuable pastures till it may constitute 80 or 90 per cent. of the sward. Other widely distributed and valuable fodder grasses are cocksfoot (*Dactylis glomerata*), rough stalked meadow grass (*Poa trivialis*), crested dogstail (*Cynosurus cristatus*), meadow fescue (*Festuca pratensis*) and meadow foxtail (*Alopecurus pratensis*). Less valuable species are Yorkshire fog (*Holcus lanatus*) and the sweet vernal grass (*Anthoxanthum odoratum*) which, though it gives a sweet scent to hay, is somewhat bitter in taste and unpalatable to cattle. Special mention must be made of timothy (*Phleum pratense*) widely distributed but some-times dominant or planted as "timothy meadows". Unlike all the pre-ceding the valuable and widespread Italian ryegrass (*L. italicum*) is not a native but was introduced from Lombardy about 1833.

Grasses themselves constitute only a part of the flora of an ordinary meadow or pasture: flowering plants have a considerable share and some, notably the white and red clovers, may have a value agricultu-rally equal to that of the grasses. Whilst some inconspicuous plants such as that favourite of cattle, the ribwort plantain (*P. lanceolata*), may have a considerable fodder value, it is unfortunately true that those which make our meadows so attractive in spring such as the buttercups and ox-eye daisies are often unwanted weeds indicative of poor management. Modern work has shown however that some previously despised "weeds" are rich in minerals essential in a balanced animal diet. Daisies dandelions, yarrow (*Achillea millefolium*) and knapweed (*Centaurea* sp.) are all important in this respect (see Plate 4a and 4b).

Although most meadow grasses are perennials and live for a number of years, they vary both as regards longevity and fighting qualities. If a nutritiously valuable species can survive in competition over a number of years, it becomes doubly valuable. There are strains and races amongst grass species just as marked as amongst other plants and animals. Seed from grass grown in a hard environment carries the fighting qualities of its parent : hence the value and popularity amongst English farmers long enjoyed by Scottish seed. Though the subject is a complex one plants grown from seed imported from warmer climates, even of the same species, may not survive.

In some respects the particular strain of grass is more important than the species. The grass should tiller well, i.e. it should send up a number of shoots : it should produce abundant soft leaves, i.e. a "leafy strain" ; it should have a long growing season so as to provide fodder over a long period. Tillering is encouraged by grazing : many farmers turn sheep out on to autumn-sown wheat so that they may both consolidate the soil by treading and manure it as well as inducing the plant to send up more branches to replace those eaten off. Different species of grass react differently to mowing and grazing so that fields regularly mown come to differ from those grazed. Further, animals are complementary grazers : cows wrap their tongues round the grass and to some extent pull, at the same time treading heavily on the roots : sheep and rabbits nibble with their teeth and cut the blades of grass ; the treading by the smaller hooves of sheep is usually beneficial.

In due course valuable grasses such as *Lolium perenne* (which needs a rest in spring and autumn) are eliminated by such a succession of ill treatment as constant grazing and treading and it is because bent (*Agrostis tenuis*) stands up best that neglected permanent pastures come to consist mainly of this species.

Where land has been ploughed, it is sometimes left to tumble to grass—a pasture of sorts is established by the invasion of native species. A second method of establishing a pasture is to sow a simple mixture such as rye grass and wild white clover to establish a close sward though actually nature providse other grasses and herbs as fillers. A third method is to sow a more or less elaborate mixture of grass seeds and clover seeds, the selection being suited to the condition of the land. Old pastures broken up and reseeded with carefully selected seeds mixtures may show a very great increase in grass or hay yield and in stock carrying capacity : they need however to be carefully grazed by securing an exact balance between grass growth and stock requirements, otherwise degeneration of the pasture, in any case always likely to happen, takes place very rapidly. All the species of grass mentioned above as valuable agriculturally (including the alien Italian rye-grass) are included in modern seeds mixtures. Others include *Poa pratensis* (smooth meadow grass), *Festuca elatior* (tall fescue), the oat grass (*Trisetum flavescens*); bents (*Agrostis tenuis* and *Agrostis stolonifera*) are rarely sown.

When Sir George Stapledon and Dr. William Davies with their team of field workers carried out the Grassland Survey of England and Wales in 1940, they divided the lowland or enclosed pastures of the country into eight groups, numbered I to VIII. In Group I, First Grade Perennial Rye grass pastures (with a sward containing 30 per cent. or more of rye grass) predominated. This Group was found only over a few restricted areas, notably in Romney Marsh and Leicestershire. At the other end of the Scale Group VIII consisted mainly of *Agrostis* pastures, with a very few of the best fields containing some rye grass but the poorer fields carrying rushes or other species characteristic of rough grazing. The surveyors found Group VIII dominant in very many parts of the country. The Survey was a practical wartime measure: for each type in each county seeds mixtures were recommended to secure rapid improvement. It must not be thought from this Survey that the ideal for the whole country is necessarily a rye grass-clover mixture. Cattle and sheep will sicken of such a rich diet and turn eagerly to "roughage": they need trace elements contained in other grasses and herbs. There are even cases, notably in the "teart pastures" of Somerset, where animals absorbed an excess of the trace element molybdenum contained in the leguminous plants of an over-lush pasture and became ill until the cause was discovered and preventive measures undertaken.

The importance of good grassland was appreciated early. The Anglo-Saxons certainly understood the significance of waterside meadows for their cattle and in providing hay and winter feed. It is no accident that early place names (ending in -ham) in Northamptonshire are concentrated on lands which Stapledon's 1940 Survey graded high in the national grassland classification. In the Middle Ages it is not too much to say that the successful manor was the one with good grassland for grazing and the provision of winter fodder. As early as the 7th century the unsurpassed Romney Marsh pastures were fully appreciated and through the centuries this famous area guarded its treasured land by the elaborate "law and custom of the Marsh". It was left to the stupidity of 1953 that a severe struggle (fortunately successful) had to be entered into to prevent a large area being covered by concrete runways for a private airfield. Ferryfield airport was established on poorer land.

This brief account of our meadows and their grasses would not be complete without reference to water meadows, once so important in the south of England, now because of labour and upkeep costs so much neglected. Especially on the alluvial flats bordering the streams passing through the chalk country, lime-rich water was used to flood meadows early in the year, the surplus water being drawn off. Not only was a good growth of early grass secured but leafy strains developed and several species of little value under normal conditions became palatable to animals and hence valuable.

It should be clear that the ordinary grass field of a British farm is far from being the simple thing it appears at first sight. It requires careful and skilful management, varying according to soil, climate and other conditions, to maintain its agricultural value. The average carrying capacity of improved farm land in Britain is about 0.4 stock unit per acre but well managed grassland will maintain at least one stock unit per acre throughout the year—i.e. one bullock or six sheep. Good meadows will mow two tons of grass per acre. Many changes are taking place : hay is giving way to dried grass, where the grass is dried immediately on being cut and then compressed to form a sort of cattle cake, with a very high protein content. Appropriate seeds mixtures of carefully selected strains are able to add enormously to Britain's ability to feed more animals and so to produce a larger home supply of meat and milk.

So far as grasses are concerned it will be noted that Britain relies almost entirely on native species, though selected ones, notably rye grass, have been consistently sown since about 1700. The grasses were there, they have been encouraged and developed. Further, Britain has given much to the rest of the world in grasses and grassland management. The prosperity of New Zealand is dependent very largely on the sown pastures of British species. If there is a region in the world which can boast primarily of its pastures it is the Blue Grass region of Kentucky in the United States. No wonder the controversy rages hot and strong : is the Blue Grass (*Poa pratensis*, the smooth meadow grass of Britain) there a native or is it just another of the gifts of Britain to the United States, dating from the early settlement ?

If the grasses of our improved pastures and meadows are nearly all native species which have been improved or developed, the same is not true of the clovers and other leguminous plants which are included in seeds mixtures or grown separately as fodder plants. Monks who practised "high farming" as early as the 12th and 13th centuries knew the value of clover in their rotation and undoubtedly stepped up the yield of wheat

FIG. 22
The Natural Regions of Britain
The Midland Coalfields should be added

from the 10 bushels per acre of the open fields to double that quantity, but it was not, as mentioned in Chapter 7, until the Improvers of the 17th century that the virtues of clover were widely extolled. From about 1700 onwards the sowing of clover—with rye grass or ray grass as it was then called—became general. Not, of course, till late in the 19th century was the work of nitrogen-fixing root-nodule bacteria, converting the nitrogen of the atmosphere into nitrogenous plant food, understood.

Of the leguminous plants entering seeds mixtures and so into pastures, the chief are White Clover (*Trifolium repens*), Wild White Clover (*T. repens, var sylvestre*), Red Clover (*T. pratense*) and Alsike Clover (*T. hybridum*). For special purposes and types of soil, more often as fodder crops, Medick (*Medicago lupulina*), Crimson Clover (*Trifolium incarnatum*), vetches or tares (*Vicia sativa*, spring and winter varieties, and other species), lucerne or alfalfa (*Medicago sativa* and hybrids with *M. falcata*), sainfoin (*Onobrychis sativa*), and lupins (*Lupinus* spp.) must be added to the list. Though authorities differ, out of these probably only white clover, medick and the common vetch are natives and even with these there has long been an import of seed from the Continent.

Sainfoin is a corruption of Saint Foin or Holy Hay and may be regarded as an important French contribution to the amelioration of the shallow dry soils of the chalk in southern England where its long roots penetrate to a great depth and reach water. The common sainfoin is usually left down for six or seven years : there is a giant sainfoin which is usually allowed to occupy the light sandy soil on which it thrives for only one year. Although so valuable on chalk lands sainfoin thrives on many types of soil and its handsome flowers add much to the beauty of southern Britain whether the plant is grown as a pure fodder crop or mixed.

Lucerne or *Alfalfa* reached Britain in the days of the Improvers as La Lucerne, but it has never been as appreciated here as in other countries. With long roots it is extremely drought-resisting and although it may take two or three years to get established, it then yields excellent fodder or hay with a high protein content for many years. To secure a good "strike" it is important to inoculate the seed with root nodule bacteria—a simple process but one which seems to frighten many farmers. A field of lucerne in bloom is one of the grand sights of an agricultural landscape—most commonly seen on light dry soils, as on the chalk.

Alsike clover is certainly one of the handsomest of the clovers and is valued as hardier than the Red Clover. Whatever its origin (probably from Sweden in 1834) the supplies of seed now used are mainly Canadian. The handsome and very common *Red Clover* has an uncertain origin, but is probably not native. It was probably introduced into Britain from the continent in the early 17th century by Sir Richard Weston —it was certainly strongly advocated by him. Like the Red Clover, the *White Clover* probably found its way to Britain from the Netherlands. Many strains of both Red and White Clovers have been developed: it is interesting that of the White a strain reintroduced from New Zealand is particularly valued.

Lupins. One commonly associates lupins with old-fashioned cottage gardens, or giving a splash of colour to a not-very-tidy herbaceous border. It seems strange sometimes to see a whole field—white, yellow, or blue-flowered—but they are grown on very light land deficient in lime and humus as a green manure. Like other *Leguminosae* the root nodules have nitrogen-fixing bacteria, hence the value to poor light land when the crop is ploughed in. Since lupins do not object to sand with an acid reaction, more use may be made of them in the future in reclamation of coastal sand dunes : at present they are rarely seen outside the eastern counties. If the plants reach maturity the seeds may be threshed like beans or peas.

PULSE-CROPS—PEAS AND BEANS

Like other members of the pea family, peas and beans have root nodules occupied by bacteria which convert atmospheric nitrogen into nitrogenous compounds. All parts of the plants are rich in nitrogen combined in a form affording valuable plant food available for the use of succeeding crops when the plants are ploughed into the ground or decay. So beans, like a clover ley, are a valuable preparation for a cereal crop. The term "pulse crop" (an older term is "black" straw crop) is used when peas and beans are grown mainly for their seed. Broadly, beans are a heavy land crop, peas flourish on lighter land. There are many varieties of field pea (*Pisum sativum*), and "garden peas" are those picked when the pea is young, green and tender, though many special garden varieties have been developed.

Both beans and peas have long played a part in British farming. A small bean, about the size of a pea, has been found associated with wheat and barley in Iron Age sites (the Celtic Bean, *Vicia faba*, var. *celtica*). Both peas and beans played a large part in mediæval farming and into the 19th century, but have fallen into disfavour of recent years. A large part of the acreage of peas is now for canning green for human consumption.

THE CROPS OF BRITAIN

Broadly speaking, clovers and rotation grasses occupy nearly 40 per cent. of the ploughed land of Britain, cereals in all rather over 40 per cent. In order the cereals are barley, wheat and oats, with—far behind—mixed corn and a very little rye. Root crops—in order potatoes, sugar-beet, turnips and swedes and mangolds—occupy about 9 per cent. of the ploughland, vegetables 3 per cent., fruit 2 per cent. This leaves some 5 per cent. for kale and other brassica crops (the cabbage group), beans, peas and minor crops, as well as bare fallow.

Taking the year 1963, barley had shot to first place in England and Wales (over 4,000,000 acres), wheat under 2,000,000 and oats (owing to decreased demand from horses) to only 600,000. Oats remained popular in Scotland. The cereals wheat, barley, oats and rye were tersely described by Fream as "grasses with enormously swollen grains used extensively for human consumption and for feeding to farm livestock". They have been cultivated for a very long time—for millennia rather than centuries—and although most people would recognize typical examples of the four, varieties have become so numerous and indeed so different from one another that separation and recognition are difficult even for the expert. The loose seed heads, or panicles, of oats remain distinctive, but bearded wheat, rye and barley can look much alike, whilst beardless barley and rye add to the confusion.

Wheat. The origin of cultivated wheat is lost in the mists of distant past. As long ago as 1866 Heer, quoted by Darwin, showed that the Neolithic inhabitants of Switzerland cultivated no less than ten cereal plants including five kinds of wheat. So ancient therefore is the differentiation of cultivated varieties that whether they were derived from the one stock *Triticum vulgare* or from distinct wild species may never be known but recent work suggests three types from three regions. The original home of the grasses from which wheats have been derived would seem to be south-west Asia, the hexaplid types (our bread wheats) Turkestan. There is abundant evidence that wheat was cultivated in Britain as early as Neolithic times. Before 3000 B.C. Neolithic invaders from the continent of Europe brought with them domesticated cattle, sheep, goats and pigs as well as wheat. The earliest remains of wheat or indeed of any cereal grains are probably those from the Neolithic Hembury Hill Fort in Devon where querns or small hand stone mills for grinding corn were also found. The wheat from these early sites, also known from impressions left when clay vessels awaiting baking were stood on floors strewn with grains, are of primitive types, each spikelet has only two grains. But by Bronze Age times, narrow pointed grains similar in form to Emmer wheat (sometimes distinguished as *Triticum dicoccum*) are also found.

Thus the evidence is clear that the small square fields of Celtic preRoman times were in part occupied by stiff-headed golden wheat, waving in the gentle breeze just as it does today. Although as noted earlier there is confusion because of the use of the word *fromentum* both for wheat and for grain generally (just as in Britain we use the word corn to embrace what Americans call the small grains) there is little doubt that what Caesar found in the rich fields of Kent and what the Britons were already exporting to the continent was wheat, despite the fact that on Bronze Age sites barley was much more in evidence than wheat.

Until the introduction of rye by the Anglo-Saxons it would seem that wheat was the main bread-grain though some barley may have been used. Throughout the centuries which followed it remained the aristocrat of the cereal world, though the poor, especially in those parts of the country where wheat would scarcely grow, had to content themselves with substitutes. This is well expressed by Harrison (1587) when he observes, "The bread throughout the land is made of such graine as the soile yieldeth, neverthelesse the gentilitie commonlie provide themselves sufficientlie of wheat for theire owne tables, whilest their household and poore neighbours in some shires are inforced to content themselves with rie, or barlie, yea and in time of dearth manie with bread made either of beans, peason, or otes, or of altogether, and some acornes among, of which scourge the poorest doe soonest tast, sith they are least able to provide themselves of better". It would seem that the poorer subjects of Queen Elizabeth I were fully familiar with the flavour of the National Loaf three and a half centuries before it was the lot of all in the Second World War. Writing in 1758 Charles Smith estimated that rather less than two-thirds of the inhabitants of England and Wales normally ate wheaten bread.

The quotation given above from Harrison is a reminder that not all parts of Britain are suitable for wheat production. He refers to soil : the limiting factor is rather climate. In broad terms the west is too wet, the north too cool in summer for the ripening of the grain. Thus in Scotland wheat cultivation is limited to the drier and sunnier eastern lowland tracts but reaches its northern limit around Moray Firth. In mediæval times if people wanted wheaten bread—which they did—they made an attempt to grow it through most of England and Wales, incidentally also through Ireland, despite indifferent chances of success. With the influx of cheap wheat from sunny lands abroad in the eighteen-seventies and onwards, the climatic and soil factors in Britain asserted their full effect and wheat cultivation became almost restricted to the drier eastern counties of England south of Yorkshire (Plate X).

Percival*, using cultivated acreage figures given by Maxey†, has estimated that towards the end of Elizabethan times in England and Wales there were $2\frac{1}{2}$ to $3\frac{1}{2}$ million acres under wheat giving an average yield of eight to ten bushels per acre. In 1771 Arthur Young's estimate‡ was 2,795,808 acres yielding 24 bushels per acre; in 1798 Billingsley§ gave 3 million acres yielding 20 bushels per acre; in 1808 Comber‖ gave 3,160,000 as his estimate; in 1846 McCulloch¶ suggested 3,800,000 acres. The first official statistics of 1866 gave over 3,275,000 plus 110,000 acres in Scotland) whilst the maximum recorded was for 1869 of 3,417,054 acres in England, 135,562 in Wales and 135,741 in Scotland. Apart from the sharp rise in 1918 there was a steady decline to a low in 1931 of 1,180,903 acres in England, 15,794 in Wales, and 5,024 in Scotland. The introduction of a wheat subsidy then produced a rise, continued during the great war effort to a peak in 1943 of 3,147,103 acres in England, 132,802 in Wales and 84,000 in Scotland.

But the British climate does not really suit wheat—by 1961 the total acreage was back below 2,000,000. Percival in the work already quoted, states very positively, "Many hundreds of kinds of wheat are grown in different parts of the world. They are classified into eleven or twelve groups or races, which are sometimes ranked as species. The wheats cultivated in Great Britain belong to two races only, namely, (1) Rivet or Cone wheat (*Triticum turgidum*, L.) and (2) Bread Wheat (*Triticum vulgare*, Host.)." Fream (1949 edition) records that about a hundred varieties of *T. vulgare* divided into winter and spring wheats are currently grown in Britain. The principal representatives of the *T.*

---

*Percival, J., *Wheat in Great Britain*, London, Duckworth, 1934.
†Maxey, E., *New Instruction of Plowing and Setting of Corne*. 1601.
‡Young, A., *Observations on the . . . . Waste Lands of Great Britain*, 1773.
§Billingsley, J., *General View of the Agriculture of Somerset*. 1798.
‖Comber, W. T., *An Enquiry into the State of National Subsistance*, 1808.
¶McCulloch, J. R., *Descriptive and Statistical Account of the British Empire, Vol. I*, 1854.

turgidum group are the bearded or awned Rivet wheats, giving a high yield but suitable only for stock and poultry feed. Amongst the bread wheats are those of good baking quality (the "strong" or "hard" wheats), and those suitable only for biscuit manufacture or stock feed— the "weak" or "soft" wheats. The better wheats are often more particular regarding soil or climatic conditions and may have a lower yield so that innumerable factors influence the farmer's choice of a variety. During the Second World War an interesting position arose: the necessity of reintroducing wheat cultivation into counties from which it had almost disappeared. The traditional derivatives of "English" wheats (such as Yeoman and Holdfast) were known to be unsuitable. It was found that varieties evolved to thrive in the cool rather moist conditions of Sweden did well in such cool counties as Northumberland and Durham (Chevalier, Crown, Iron III, Steel, Scandia and Weibull's Standard). On the other hand Dutch varieties did extremely well in the dry eastern counties where conditions agree most nearly with those in the Netherlands (Wilhemina, Juliana). Some French wheats (such as Desprez 80 or Joncquois) did very well in the south. The ones mentioned are all winter wheats; there is a similar differentiation amongst spring wheats.

It is indicative of the remarkable changes of recent years that in the brief period from 1939 to 1945 the twenty-six most popular wheats completely changed their relative positions and thirteen of them are not even mentioned by Percival writing in 1933. By 1950-51 one third of all the wheats sent to the National Institute of Agricultural Botany were Atle and Bersea—both recorded separately for the first time only in 1942-43. These kaleidoscopic changes have continued. By 1956-57 Cappelle Duprez (29 per cent.) led the winter wheats, Koga II (20 per cent.) the spring varieties, both unknown a few years earlier.

That each type of land has varieties of wheat best suited to it has long been appreciated by the practical farmer. As William Ellis wrote over two centuries ago, "Old Red Lammas has a red Straw, and a red Ear; this is reckoned the best of Wheat because it makes the finest Flower. It answers best on rich Lands. Yellow Lammas has a red Ear and white Straw, and is reckoned the second best. Pirky Wheat is the most convenient for our Chiltern Lands. Dugdale Wheat has a four-square Ear, is a hardy wheat, will grow on sour Tilth the best of any. These four sorts are what they chiefly sow in *Hertfordshire*." None of these varieties survive into the present day but they are the parents of the many improved varieties of British wheats.

Barley. Barley is almost as ancient a grain in cultivation as wheat. Six-rowed barley was certainly known to Neolithic settlers in Britain.   In Iron Age settlements wheat, barley, oats and beans were all cultivated and grains of barley (including two-rowed types) and oats associated with wheat have been found in many Roman and Romano-British settlements.   Although barley may have entered into bread in early times as it has done at intervals or in a subordinate way since, it is broadly true to say that wheat was the bread grain, barley the drink-grain.   The malting of barley is certainly very old in Britain.

Although all the numerous varieties of barley are probably derived from *Hordeum vulgare,* two main "species" are now of importance to the British farmer : the two-rowed barley sown in spring (*Hordeum distichum*) and the six-rowed barley (*H. polystichum* or *H. hexastichum*). Doubtless in the old days of British farming barley was sown as widely in extent over the country, though not covering the same acreage, as wheat.   In modern times with world wide competition it really only pays to grow barley where natural conditions are favourable.   It is even less tolerant of excessive moisture than wheat and so is a grain of the drier, sunnier eastern counties specially if grown for malting.   It is able however to take advantage of long summer days and so grows farther north than wheat.   Indeed Arctic barley ripens well within the Arctic Circle. Barley is a shallow rooting crop of light soils whereas the heavy headed wheat does best on heavier land.   Normally spring sown, barley can be used where a winter sown crop of wheat has failed.   The best barleys are used for malting : where the grain does not reach the requisite standard it is excellent stock feed.   So barley, in many ways, is nowadays a complementary crop to wheat rather than a rival.

Rye. Rye is well described as the "poor relation of wheat".   It was a much later introduction : it was doubtless brought to Britain by the Anglo-Saxons.   To this day a favoured, even a preferred, bread grain in continental Europe, it has never been appreciated to the same extent in Britain.   Broadly it prefers the conditions of soil and climate suited to wheat ; being a poor relation it does not get them and is relegated to poorer, thinner and more acid soils, where wheat would not survive. The resuscitation in cultivation during the Second World War was for the purpose of using, for a bread grain, land which would not grow any other bread cereal.

Oats. Although oats were probably derived in early times from several species, the cultivated strains grown in Britain are probably all derived from *Avena sativa.* This is believed by some to be but the cultivated form of the native wild oat, the pleasantly named *Avena fatua.* Whether that is correct or not, the damp-tolerant oat is preeminently the one grain crop for all Britain. Though not strictly a bread grain it has played a large part in human food, especially in Scotland, as oatmeal, oatcakes and porridge. The last serves still as the mainstay of diet in many a Scottish home. Elsewhere oats of several varieties serve as animal feed : if the season does not permit ripening the crop can be cut green, dried or made into silage.

In the past various cereal mixtures have been grown. Maslin, or mixtelyn, miscelin or mestlen, was a mixture of wheat and rye. Especially since about 1918 "dredge corn" has become a popular fodder crop in the south-western counties or England—it is a mixture of oats and barley and the yield is claimed to be greater per acre than if either were grown alone : further if the season causes the failure of one the other will survive and thrive.

America's great gift to the world's agriculture, maize (*Zea mais*) is not really a farm crop in Britain. In sheltered localities in the eastern counties good cobs of sweet corn can be produced at least in some seasons and command good prices as a fresh vegetable.

Buckwheat (*Fagopyrum esculentum* formerly *Polygonum fagopyrum*) may, from evidence of seeds, have been important in early days, but its position is uncertain. Was it a weed with a useful seed ? Was its seed gathered and used for flour as it is in some countries today ? True it will grow on light sandy soils where little else will survive and can be sown after another crop has failed. At present it is seen in patches in East Anglia—grown where pheasants may come and help themselves.

Root Crops

If one ventures to use the term "agricultural revolution" it is almost synonymous with the introduction of root crops. They solved the age-old problem of winter feed. There was no longer the need for the autumn slaughterings, the salting of meat for winter use. The struggle had always been to keep alive through the winter enough stock for breeding by hay from the waterside meadows. Sown in the summer, "lifted" in autumn, stored with little difficulty in "clamps" to be used through the winter, turnips, Swedish turnips or "Swedes", and mangolds or mangel-wurzels

ARABLE

O MILES 50

FIGS. 23 (*above*) and 24 (*facing*)
Arable Land in Britain, 1930-1933
The solid black is the arable land and the concentration in the drier east is very
marked. Reduced from the detailed maps of The Land Utilisation Survey

ARABLE

O MILES 50

completely revolutionized stock farming. Almost as far reaching has been the influence of the potato, now the universal vegetable for human consumption, the vegetable that made Ireland into a land inhabitable and inhabited by 8,000,000 people and then equally broke Ireland and reduced the population to 4,000,000. The sugar-beet was Napoleon Buonaparte's chief weapon in breaking the British world monopoly in sugar, and for a century and a half it has been a keystone in the economy of half the countries of Europe. Our early ancestors knew nothing of any of these root crops, and our forebears were incredibly slow to realize their advantages and to adopt their wide cultivation.

The potato had long been in use in South America when Columbus discovered the New World in 1492. It was described in Gerard's *Herball* in 1597 but he probably thought the potato and sweet potato were identical. The true potato was probably introduced into England by settlers who returned from Virginia in 1586. Sir Walter Raleigh grew it on his estate in Ireland. The new Royal Society set up a committee in 1662 to look into its possibilities (report issued 1663) and it took a firm hold in Ireland and was cultivated in Lancashire where Wigan had a potato market by 1690. Advocated for large scale farm cultivation by Worlidge in 1699, it remained a garden and not a field crop for another century. Broadly, as Trow-Smith points out, the cultivation of the potato kept pace simply with the demand for a cheap food by the growing industrial population. So the potato took about a couple of centuries to establish itself in Britain. Undoubtedly the potato is the greatest contribution which has been made by the New World to the food economy of the Old—even greater than maize. Cultivation and consumption have become universal. When the potato was discovered by the Spanish explorers of Central and South America it had already a long history of domestication and adaptation. The simple wild potato had been developed into the species complex known as *Solanum andigenum* over the high Andes of Bolivia, Peru and Colombia. Thence vigorous, high-yielding strains had been taken to the lowlands of southern Chile and adapted to conditions closely comparable to those of north-west Europe. It was from this area that they were first brought to Europe.

Strange as it may seem the turnip had been known in England many

---

Plate XIII. BRITISH DAIRY CATTLE.
a. (*above left*) Jersey. *b.* (*above right*) Guernsey. *c.* (*below*) Friesian (Royal Show Champion 1951). (See page 121). (*Farmer & Stock-Breeder Photographs*)

centuries as a garden vegetable before it was first seriously advanced as a fodder crop by Sir Richard Weston, one of the great "improvers" in 1645, and eulogised by Speed in 1659. Its advocacy by Viscount Townshend—Turnip Townshend—is too well known to be more than mentioned, but the period was 1730-1738, nearly a century after Weston. For all that time the key to the winter hungry season was lying idle, waiting to be used. It was of course the use of turnips which made possible the successful breeding work of the Colling brothers and others in the 18th century. Swedish turnips, now universally known as swedes, have a longer period of growth, are hardier, have a large percentage of dry matter and store better. They were introduced in the latter part of the 18th century.

The mangold or mangel-wurzel, was probably introduced into England from France in 1786. It is said that at first only the foliage was used and the roots thrown away until by accident it was noticed that swine relished them. The crop was popularized by J. C. Lettsom by an advocacy which matched that of Townshend for the turnip half a century earlier. Mangolds give a much higher yield of dry matter per acre than swedes or turnips but have more exacting climatic requirements— demanding the warmer summer temperatures of the south and south-east of England.

In a way the three fodder roots have served their purpose, have had their day and are on the way out. They are giving place in the farming economy to sugar-beet, which has a dual purpose; to leafy varieties of the brassica or cabbage group (especially kale) as fresh winter fodder; and to such winter keep as dried grass and silage.

Today the emphasis is thus increasing on such derivatives from the cabbage (*Brassica oleracea*) as Marrowstem kale, Thousand-head kale, Rape kale, Hungry Gap kale, and Kohl-rabi, as well as cabbages themselves. This is particularly interesting, since, as already pointed out, the cabbage is one of the oldest of vegetables. Thus the solution to the problem of winter keep has actually been in the hands of the farmer, though he did not know it, for at least two thousand years.

Sugar-Beet. The story of sugar-beet in British agriculture is a curious one. In its wild state it is a native of sandy shores of southern Europe,

---

Plate XIV. A PORTRAIT GALLERY OF SHEEP.
a. (*top left*) Cheviot. *b.* (*top right*) Border Leicester. *c.* (*middle left*) Leicester. *d.* (*middle right*) Lincoln Ram. *e.* (*bottom left*) Southdown. *f.* (*bottom right*) Suffolk Down Ram. (See page 139). (*Farmer & Stock-Breeder Photographs*)

with a thin tap root, but one very rich in sugar. In the cultivated plant the tap root is enlarged, the content of sugar in the juice maintained or increased to an average of over 20 per cent. In the early part of the 19th century when the world was dependent on cane sugar from the tropics, Napoleon saw the possibility of making his European Empire self-sufficient in sugar, and so safe even with British control of the seas. The policy he initiated, despite early setbacks, was eventually so successful that sugar-beet became a leading crop in many European countries. By 1886 roughly half the sugar produced in the world was beet-sugar. By 1900 Britain was buying huge quantities from the continent, two-thirds of her total consumption, but none was produced at home : indeed sugar-beet was unknown as a farm crop. Nearly half of all Britain's sugar supply was coming from Germany when the First World War broke out, and efforts which had been made since 1909 to start cultivation of the beet and production of sugar (including the building of a factory largely with Dutch capital in 1912) in England came to fruition with Treasury help. Since then sugar-beet has become a familiar crop wherever land is reasonably accessible from one of the 17 factories. The crop takes the position of a root crop in a rotation ; beet lifting takes place late (October-December) so that it interferes little with other farm work, the tops are cut off and furnish cattle feed, the pulp returned from the factories to the farms is valuable winter keep.

It is not possible here to deal with all farm crops. Brief mention should be made of those which make their conspicuous marks on the countryside. Amongst these are the bright yellow fields of mustard in flower : white mustard for sheep feed or green manuring, both black and white mustard for seed and the manufacture of table mustard and mustard oil. Mustard is a crop of the rich soils of eastern England. Rape is a forage crop, especially for sheep, looking rather like a swede, except that no bulb is formed. A field of flax in flower is another sight to be remembered: the lovely pale blue flowers show from afar as a bluish green sheen. The time when a flax-patch was part of every farm garden has long since gone—with the old sources of dye, madder and woad, though a little saffron is still cultivated. Only the older remember the saffron cakes as one of their childhood delights.

CHAPTER 11

EVOLUTION IN THE FARMYARD*

ACCORDING TO RUDYARD KIPLING, whom we do not necessarily accept as a final authority, "The Dog was wild, and the Horse was wild, and the Cow was wild, and the Sheep was wild, and the Pig was wild—as wild as wild could be—but the wildest of all the wild animals was the Cat. He walked by himself and all places were alike to him." The dog was the first to be tempted by the smell of roast wild mutton coming from the mouth of the cave, and so bartered his freedom to become guardian of the home and the helper of man the hunter. Wild horse succumbed next, to the lure of fresh cut grass dried by the fire—showing that dried grass is not a twentieth century invention. Next wild cow promised daily milk in exchange for the wonderful grass from the meadows. One presumes that wild sheep and wild pig, brought in by man on his new friend the horse, likewise succumbed to the prospect of a constant food supply. Only the cat bargained successfully for partial liberty.

The picture remains broadly true. Perhaps the cat, at least in some of the many forms developed by man for his amusement, has become more dependent upon a purely urban environment and walks less by himself. The domestic cat population is largely guesswork but in Britain was estimated by P.E.P. (Political and Economic Planning) in 1957 at 5,200,000 against 3,700,000 pet dogs and 6,000,000 cage birds—in all costing £50m. a year to feed. The cat impact on the rodent population, especially on the black rat (*Rattus rattus*), the ubiquitous brown rat (*Rattus norvegicus*) and the domestic house mouse (*Mus musculus*) is considerable and beneficial and on the bird population is also considerable but mainly deplorable. If a cat on an average drinks a quarter of a pint of milk a day, then a herd of 100,000 good milking cows is kept fully employed in providing the necessary supplies!! Perhaps the meat and fish consumed would otherwise be largely wasted but in total must be of the order of 25,000 tons a year.

*I am indebted to my fellow editor James Fisher for comments on this chapter.

The number of dogs is more accurately known, since a dog of over six month must be licensed. In 1949-50 Dog Licences issued numbered 3,089,450 (2,890,505 England and Wales, 198,945 in Scotland). Since dogs kept solely for the purpose of tending sheep or cattle on a farm, as well as dogs kept by blind persons for their guidance, are exempt, we may be certain that dogs and puppies at any one time exceed 4,000,000. Broadly one household in five keeps a dog.

Carl Sauer in his recent review of the evidence follows Studer and Dahr and accepts the monophyletic origin of the dog with the original hearth or home and area of first domestication in south-east Asia. A recently extinct wild dog has been described from Java, and a Pleistocene wild dog from caves near Peking. The semi-wild pariah or 'pie' dogs of India and Indo-China and the Australian dingo may be accepted as very close in character to the primitive dog. Sauer believes that the dog became a household or domestic pet in the first instance and that utilitarian aspects as in the chase or as guardian of the home and flocks were not recognized till much later. The dog does not seem to have been known to Palaeolithic man but was introduced with the wave of Mesolithic culture from the east and was firmly established in Britain as the companion of Neolithic or cave-man. The cat was a much later introduction into Britain—probably in late Anglo-Saxon times.

CATTLE

Even the most casual observer, travelling through the farmlands of Britain, can scarcely fail to be struck by the number and distinctive character of the breeds of cattle. In recent years the British Friesian, black and white, has become the ubiquitous dairy cow of the Lowlands and has largely replaced the Shorthorn with its confusingly wide range of colouring. With other breeds of cattle colouring is often the easiest guide to recognition. Within the post-war years dehorning has become so general that horned cattle are almost a rarity.

On the fattening pastures of the Midlands—for example in Leicestershire—one may often see individuals of five or six of the chief beef and dual purpose breeds in the same field, for there it is the business of the grazier to buy 'store' cattle from many parts of the country and to fatten them for the butcher. Elsewhere in the lowlands of Britain dairy herds may differ from one farm to another according to the preference—often strongly held—of the individual farmers and it is not difficult in a radius of fifty miles from London to find dairy herds of each of the dairy breeds as well as of many dual-purpose.

Nevertheless, over a large part of Britain each area still has its characteristic breed or breeds of cattle. In some cases the attachment to a particular district is extraordinarily strong. Thus over North Devon the sturdy Red Devons remain popular: it is possible to draw a sharp line across the country to the south of which supremacy passes to the lighter coloured South Hams. The curly coated broad white-faced Hereford cattle still dominate the rural scene in the county from which

BREEDS OF CATTLE

	Bulls Licensed		Entries at Royal Show			
	1937-38	1953	1889	1939	1953	1963
BEEF BREEDS						
Aberdeen-Angus	755	518	87	181	69	88
Devon (two types) ..	1,091	305	84	36	29	28
Galloway	381	132	46	77	53	59
Hereford	2,116	1,913	128	63	97	145
Highland	7	2	121	20	19	13
Sussex	235	149	97	29	25	26
DUAL-PURPOSE BREEDS:						
Shorthorns	23,730	5,818	222	253	238	145
Dexter	33	40	59	45	30	17
Lincoln Red	1,324	656	—	55	65	25
Red Poll	569	358	71	160	111	77
South Devon	439	257	—	31	22	c
Welsh Black	349	257	49	19	14	12
DAIRY BREEDS:						
Ayrshire	489	2,203	50	131	215	85
British Friesian	2,671	6,432	—	207	166	253
Guernsey	1,924	1,074	141	175	89	97
Jersey	545	795	434	222	117	185
Kerry	19	8	77	15	13	c

Data from F. H. Garner, *The Cattle of Britain* (Longmans, 1944) and figures for 1953 supplied by Ministry of Agriculture. The Royal Show now only separates Beef Breeds (as above plus Beef shorthorns (38 in 1963), Lincoln Red, Welsh Black and Longhorns (11 in 1963)) and Dairy-Dual Purpose (as listed above for Dairy plus Dairy Shorthorns, Red Poll, Dexter, Red-and-White Friesians (11 in 1963)). c=classes cancelled. In recent years Charolais bulls have been introduced from France to improve meat-yield from dairy breeds.

they take their name : similarly the Galloways seem almost to exclude
other breeds in parts of the Scottish district of Galloway. If the popular
dairy breed the Ayrshire is now found in many parts of Britain it is still
true to say that it reigns supreme in Ayrshire itself.

The facts of distribution of cattle breeds may be observed but
we know extraordinarily little of the reasons behind the distribution. A
farmer will often claim that no other breed but the one he has chosen will
flourish on his farm, that others have been tried and failed. If such
assertions can be substantiated there would seem to exist some very
subtle differences in the environmental factors of the cattle habitats in
Britain—differences so subtle that they have not yet been determined or
isolated. Can it really be true that Herefords will flourish on the Welsh
borderlands but not everywhere in Britain when, taken abroad, they
seem not only to survive but to thrive on the arid sun-baked plains of
Texas ? It has been claimed that Highland cattle transported from the
Scottish Highlands to the alien environment of Exmoor went berserk
and had to be shot whereas on the wide open ranges of South Dakota
they become the kindliest and gentlest of beasts whether the temperature is 30°F. below zero or 116°F. above.

A few years ago under my direction Miss Joan O'Connor attempted
to trace the influence of environmental factors on the current distribution of cattle breeds in Britain. Her work was of extraordinary interest
and has been freely used in the paragraphs which follow. It must be
admitted however that three years of intensive work failed to reveal any
features of the natural environment exercising a decisive role. In the
buying and selling which takes place between farmers at the local
market, in the immediate availability of bulls for service, in the production of a certain standard and type of milk there is an obvious
advantage if the cattle in a given area are of one or two main breeds.
Such considerations would seem to exercise a much greater influence on
distribution of breeds than any natural environmental factors.

If it is difficult to determine the reason for the present distribution of
the different breeds of cattle in Britain, it is almost equally difficult to
determine their origins. For each of the principal breeds there is a
Breed Society : conditions are laid down for the registration of animals
and the Society publishes a Herd Book—in annual or periodic volumes.
With due precautions (since the name and address of the owner does not
guarantee that the registered bull or the herd lives at the same address)
the various Herd Books enable the distribution of the breeds to be
mapped and changes in popularity and distribution to be traced over

the years. When a variety has become sufficiently established for it to be recognised as a distinct breed a new Society and Herd Book may be set up. Once this has been done a certain element of stability is apparent—the purity and distinctive character of breeds are recognized and the many mongrels omitted. Many breed societies have been in existence for well over a century ; most have at some time or another published accounts of the origin of the breed though in the majority of cases detailed factual evidence of unimpeachable character is extremely meagre.

ARTIFICIAL INSEMINATION, ENGLAND AND WALES*

Breed of Bull			Number 30 Sept. 1953	First Inseminations 12 months to 30 Sept. 1953.	
British Friesian	..	..	212	444,880	37.8 per cent
Shorthorn ..	..	..	192	206,795	17.6
Ayrshire	..	..	122	131,094	11.1
Hereford	..	..	63	129,694	11.0
Guernsey	..	..	61	91,198	7.8
Jersey	..	..	45	49,868	4.2
Devon	..	..	18	41,057	3.5
South Devon	..	..	12	30,166	2.5
Aberdeen-Angus	..	..	20	19,608	1.7
Lincoln Red	..	..	6	11,935	1.0
Red Poll ..	..	..	9	9,197	0.8
Welsh Black	..	..	9	6,860	0.6
Sussex	..	..	4	4,052	0.3
Galloway ..	..	..	4	724	0.1
		Total	777	1,177,128	100.0

A little over a couple of centuries ago cattle in Britain occupied a position not unlike that which they still occupy amongst some of the pastoral-agricultural peoples of the tropics. A young village Burman will scarcely be regarded with favour as a prospective son-in-law until he has acquired a pair of bullocks. He then has the means to plough the rice field or millet-patch and to drag to market the two-wheeled cart. Sturdy muscular animals are called for. In Britain of old oxen were the plough animals and even as late as 1840 some writers extolled their virtues for this purpose as vastly superior to the horse. It was even argued that in the deep mud of the trackways and on heavy clay lands long-legged cattle had an advantage and were to be valued accordingly. It

* In Scotland over the same period 43 bulls gave 49,909 first inseminations, 61 per cent Ayrshire, 17 per cent Friesian, 8 per cent Aberdeen.

was not indeed until the latter half of the 18th century that the increasing demand for meat from the towns created by the Industrial Revolution led to attention being seriously given to the breeding of cattle primarily for meat. Later still came the development of the dairy breeds and attention to quality and quantity of milk. Even when meat and milk became the chief objectives of cattle breeding and rearing the emphasis continued to change. There would seem to be a certain obvious advantage in a breed yielding both good meat and good abundant milk—in other words in a dual-purpose breed with bullocks for beef and cows for milk. Many farmers still pin their faith to dual purpose animals : they argue further that there is a great advantage when a cow no longer yielding sufficient milk to be an efficient member of the milking herd can still be fattened before slaughter to yield reasonable quality meat. On the whole however the modern trend is towards specialization and not to confuse meat and milk. Although the Shorthorn is classed in the table given above as a dual purpose animal most Shorthorn herds are either dairy herds or beef herds.

Breeds continue to rise and fall in popularity with changes in public taste or economic circumstances. So long as cattle were plough animals it was bone and muscle which mattered. When the emphasis changed to meat the need was for a well-knit animal with small bones in proportion to flesh. For long the roast beef of England was designed to form the Sunday joint of a family of perhaps a dozen children as well as numerous servants. Large joints were in favour and a goodly proportion of fat was regarded as a sign of quality. In the present century smaller families led to a demand for small lean joints. Instead of large animals fed to fatness and killed at four years, the farmer was led to specialize in 'baby beef' smaller animals ready for the butcher at two years, since the British taste has never been primarily for veal.

The yield of milk varies widely from breed to breed as well as between individuals. There are records of Friesian and Ayrshire cows yielding over 3,000 gallons of milk in a year, or over 8 gallons a day, but a 'good milker' would apply to any yield over 1,000 gallons. The average for a good Shorthorn herd would be between 700 and 850 gallons, or double what is commonly given as the 'average' for cattle throughout the country. Actually this 'average' gives a false picture since it includes the dry cows of a dairy herd : the average of cows in milk is more like

---

Plate XVa. (*above*) Kent or Romney Marsh (page 146)
b. (*below*) Lincoln Yearling Ewe. (page 146) (*Farmer & Stock-Breeder Photographs*)

Plate XVIa.—Wessex Saddleback (*p. 150*)

b.—Large White Sow and litter. Chivers & Sons, Histon, Cambridge (*p. 149*)

600 to 650 gallons. Very high yields in milk tend to be associated rather naturally with poor quality—low in butter-fat and low in salts—and it has become necessary to safeguard the public by imposing minimum standards. The Jersey cow has achieved a special place in public esteem by its yield (only moderate in quantity) of a milk which by its darker colour has a rich appearance as well as in fact having a high butter-fat content and yielding much cream. Such milk however may be so rich as to be more difficult to digest than one in which the butter fat content is lower and in very small globules.

In the years between the two World Wars the Government through the Milk Marketing Board gave the farmer a guaranteed market and a guaranteed price for milk. Dairy farming was so stimulated that milk was produced in excess of demand and campaigns were launched urging the public to "drink more milk". The maintenance of home milk supplies during the Second World War was a major factor in the national health, especially of children, and expanding production led to all demands being met after the war despite continued low yields of some areas. It must not be forgotten that there are still many parts of the country where the mixed farmer keeps a few cows to supply milk for himself and family, makes a little cream, feeds skimmed milk to his pigs, and so is content to use any old bull for breeding.

The origin of British cattle, like that of most of our domestic animals, is not known with certainty. In Pleistocene times there were at least two species of wild cattle, *Bos primigenius* and *Bos longifrons*. Skulls of *Bos longifrons* have been found with flints embedded in their foreheads, suggesting that they were hunted by Stone Age man and it is generally agreed that this species was also domesticated, and is the origin of many British and continental breeds. They were moderate-sized animals with short thick-set heads, short bodies with fine bones and small horns. Some writers have claimed they were not unlike the Irish Kerry of today. Although it has been stated that the pre-Roman British cattle were of various colours such as yellowish-grey, there is an interesting theory that they were mainly if not entirely black and when later displaced over the lowlands of Britain they became associated with Celtic Britain. It is certainly true that the present distribution of the black breeds—Aberdeen-Angus, Galloways, Kerries, Dexters, Welsh Black and Gloucesters —is closely associated with the Celtic fringe.

By way of contrast the wild *Bos primigenius* was a gigantic animal, larger than any domesticated cattle of the present day, and with huge horns of great span exceeding that of Highland cattle. It may well be that in days when strength as a plough animal was the quality most to be desired *Bos primigenius* attracted the hope of primitive man. It has been claimed that the Longhorns, Scottish Highland and Park Cattle were derived primarily from the domestication of *Bos primigenius*. It has been presumed that these large animals were also both very wild and fierce and surprise has been expressed that such should be domesticated by man. Some exceedingly wild and dangerous animals, notably the *saing* of Burma or wild buffalo which is almost indistinguishable from the domestic buffalo become so docile in domestication as to obey the least command of a naked boy of five or six though, incidentally, retaining a hatred of Europeans. The same may have happened with *Bos primigenius*.

From Roman sources it is known that the Britons at the time of the Roman invasion had large numbers of cattle and that they used both the flesh and milk. The Romans undoubtedly introduced other cattle and may have initiated the Britons into the art of cheese making. Incidentally the Romans learnt the art of butter making from the Gauls and Germans, using butter as a salve for wounds received in battle. A comparison with Italian breeds has suggested that the cattle introduced by the Roman were white and may have been (in contrast to the view mentioned above) the origin of the old English Park cattle.

The Anglo-Saxon invaders brought with them the cattle for their eight-ox plough teams : probably red cattle similar to those still found in western Germany. It is certainly interesting that red breeds today are found south and east of the black and old white breeds as if the older cattle inhabitants were displaced on the lowlands by the new invaders. The red breeds include the Lincolns, Herefords, Devons, Sussex and Red Polls (or Norfolks and Suffolks).

It would seem that the Scandinavan invaders were responsible for the introduction of hornless or polled breeds to all those parts of the coasts which they penetrated. Many of the local breeds have become extinct but, to quote F. H. Garner, "The breeds that were developed from this movement seem to be Suffolk Duns (now Red Polls), Northern and Yorkshire Polls, or Polled Teeswater cattle (now extinct, but this may account for the polled Shorthorns that breed true in America), Angus Doddies and Buchan Humbles (now Aberdeen-Angus), Sutherland Polls (extinct)—Skye Polls (extinct), the Galloways, Devon Natts (extinct, but formerly found in South Devon), and Somerset Polls (extinct). In many instances these cattle showed brindled and dun colours."

The Normans were military conquerors, not farmers, and for centuries it is probable that no new introductions took place. Why during that long period should so many distinct breeds have remained isolated in Britain ? We have pointed out how difficult it is to suggest environmental factors and that there seems little support for Garner's contention that "There can be no doubt that the present localized homes of the majority of the breeds are ideal for them because the stock breed regularly and maintain their breed characteristics under their respective environments." If this is the case, why should the different breeds flourish abroad under entirely different environmental conditions ?

It would seem rather that the bad communications which existed for so many centuries resulted in an isolation of areas so that there was little interbreeding between one region and another. When the time came it was mature animals for slaughter or store cattle for fattening which were sold off—not breeding stock.

In those middle ages before enclosure oxen were valued for their powers of draught, cows for their milk. It is probable that very little butcher's meat was eaten and that must have been of the toughest and poorest description. The difficulty was the absence of winter keep. There was hay and straw but no swedes, turnips or other roots. The best period for feeding cattle was after the hay harvest when worn out oxen and aged cows were turned out on the aftermath and, later, on the stubble of the corn fields. Indifferently fat, they were then killed and eaten while fresh or salted for winter consumption. The younger animals had to be kept somehow through the winter on hay, straw and tree cuttings, with what they could pick up on the overstocked common grazings. Calving in the spring resulted in nearly all milk being available from about April to September, though even then as now autumn calving was made a definite objective because of the great value of winter milk.

Gervase Markham, a voluminous writer on matters agricultural and whose many books were published between 1593 and 1635, gives us a general picture of the cattle of his time. The long-horned cattle of Yorkshire, Derbyshire, Lancashire, and Staffordshire he records as the best for meat ; the tall, long-legged Lincolns, pied but generally with more white than colour, he regarded as best for labour and draught. The cattle of Somerset and Gloucestershire (in which we see the origin of the Devon) were generally blood red in colour but shaped like the Lincolns.

J. Mortimer's *Whole Art of Husbandry* was published in 1707 and from him we get a picture of the cattle breeds before the great breeding experiments of the later half of the century. Daniel Defoe, better known perhaps to the general public as the author of *Robinson Crusoe*, was an indefatigable traveller and the first volume of *A Tour Through the Whole Island of Great Britain* was published in 1724 and provides much corroborative information. There were the Longhorns in the north (probably Defoe's 'large noble breed of Yorkshire'), the upland Black cattle of Wales, and the black cattle of Angus and Galloway driven southwards in thousands every year for the new industrial towns. Above all in interest, were the "long-legged short-horned Dutch breed" of the east which awaited the attention of the Colling brothers.

Robert Bakewell (1725-1795) has so often been lauded as the founder of the real art of animal breeding that it is a little difficult to assess his real place in retrospect. In the words of Lord Ernle, "A time was rapidly approaching when beef and mutton were to be more necessary than power of draught or fineness of wool. Bakewell was the agricultural opportunist, who saw the impending change and knew how it should be met. By providing meat for the million, he contributed as much to the wealth of the country as Arkwright or Watt."*

He was barely twenty when he started his breeding experiments and in 1760, at the age of 35, he succeeded to the sole management of his father's farm at Dishley, Leicestershire. There he was visited in 1770 by that intrepid traveller and agricultural writer, Arthur Young, whose vigorous and picturesque prose painted a word picture of a florid stocky farmer in scarlet waistcoat and leather breeches who might have been the prototype of John Bull and whose lavish hospitality had become as well known as his success as a breeder. It seems however that he later became bankrupt and died in poverty, though accounts are conflicting. His main work was with sheep—the evolution of the new Leicesters— and he was less successful with cattle where he attempted to establish an improved breed of Longhorns.

Bakewell's simple principle was in-breeding. Instead of crossing one breed with another as others had done, he selected animals of the same breed, indeed of the same family, using for breeding purposes those which exhibited such features as smallness of bone and which were best in the most valued meat joints which he wished to accentuate.

Just as breeders of dogs at the present day seek to establish and accentuate certain points of form or colour which have been arbitrarily laid down as requisite or desirable, so previous cattle judges had stressed such non-essential features as shape and curve of horns, shapeliness of head, neck or legs. A cattle show was a bovine beauty contest from

*English Farming, Past and Present. New Ed. 1936, p.181.

F. Fraser Darling

Plate 5. HIGHLAND CATTLE

In their natural environment of a highland glen, Morvern, Argyll   (*p. 132*)

John Markham

Plate 6. Aberdeen Angus

The typical beef cattle of eastern Scotland are black hornless animals (*p. 198*)

which mundane utilitarian considerations were eliminated. Bakewell changed all this and sought to combine beauty with utility, looking especially for the rapid development of good flesh with adequate fat. His actual methods he kept secret but his success was such that his basically simple idea swept the farming community and has of course remained the principle of selective breeding amongst all domestic animals (except man) to the present day. With cattle his selection of the Longhorns was unfortunate, and it may be said that though he established the principle he failed with his cattle. His great success was his sheep. The Longhorns were slow-maturing and it remained for one of Bakewell's pupils, Robert Colling, and his brother Charles, to establish the most successful of all British breeds, the Shorthorn. Charles Colling established the Kelton herd of dual purpose Durham Shorthorns in 1784 and his bull "Hubback 319" became one of the most famed animals of all time, as did its descendant the Durham Ox which was taken by specially constructed carriage and exhibited throughout England from 1801 to 1810. Hubback 319, the "father of the improved Shorthorns" was actually purchased by Robert Colling on the good quality of his calves. This is perhaps the first example of the "proven sire", now a universal criterion of worth in a bull.

In the 18th century both cattle and sheep were thus enormously improved. They matured more quickly : the weight of meat increased in proportion to weight of bone just as did yield of milk in cows. Although it is difficult to get reliable figures by which to compare animals before and after the work of Bakewell and the Colling brothers, it is clear that great progress had been made. However, size and weight have long since ceased to be the final criteria of success in cattle or sheep breeding. The demand is for the small animal, with fine textured rather than coarse meat and with the fat finely distributed through the lean rather than in solid masses around it. Rapidity of growth is vital, but the product must be one of good flavour. Mere comparisons of past and present sizes and weights thus have little importance.

The Durham ox was reported to have weighed 34 cwt. (3,808 lbs.) in his prime; weights up to 36 cwt. (4,032 lbs.) have been recorded for Sussex cattle but anything exceeding a ton (2,240 lbs.) is large even for Herefords and Devons, both heavy cattle. Two hundred years ago cattle slaughtered in the London market however only averaged 900 lb. (550 lb. dead weight), but the animals were five or six years old compared with 2-2½ years in the period before the First World War. The change has been in quality rather than size and weight.

M.L.—K

Some time during the Middle Ages there were importations of cattle from Holland but, that apart, the changes were due to the breeding of British stock. It was not till the nineteenth century that there came the very significant importation of the Jersey, Guernsey and Friesian to which reference is made below.

Some notes may now be given on the characteristic breeds of cattle to be seen today in the British countryside, forming indeed one of the most conspicuous features of the cultural landscape. The distinction into Beef, Dairy and Dual-purpose breeds is a modern and largely artificial one which indeed breaks down where several breeds, including the Shorthorns, are concerned. Colour forms a better guide to breed relationships.

The Black Breeds

The *Welsh Black* is certainly one of the oldest breeds in the British Isles and the modern breed, which is the result of the merging of several local types (notably the Anglesey of North Wales and the Castle Martin or Pembroke of South Wales), is probably the direct descendant through several lines of the British cattle from *longifrons* stock of pre-Roman days. The animals are extremely hardy : cows give yields of milk under conditions in which cows of many other breeds could scarcely survive, whilst the steers (known to graziers who buy them for fattening on Midland pastures as Welsh Runts) are valued because they can be fattened on wind-swept marshlands or acid hill-pastures alike. Local environments in Wales give rise still to local variations in type and in milk yields from the cows (Plate 7a).

The Irish *Kerry*, indigenous in the south and west of Ireland is smaller than the Welsh Black but is likewise able to thrive on the poorest of herbage and in rigorous exposed climatic conditions. They too are entirely black and are believed to be direct descendants of the early British types. They have been developed as dairy cattle, though relatively unimportant. The *Dexter* may be regarded as a freaklike offshoot of the Kerry and are still smaller animals with very short legs. About 25 per cent. of all calves born to pure Dexters are monstrosities which are either premature or survive birth by a few days only. If Dexters and Kerrys are mated, however, the calves are normal. Though commonly wholly black, wholly red Dexters also occur.

The *Galloway* takes its name from the old Scottish province of Galloway, now the counties of Wigtown and Kirkcudbright, and is characteristic of the whole tract from Wigtown to Cumberland. Like

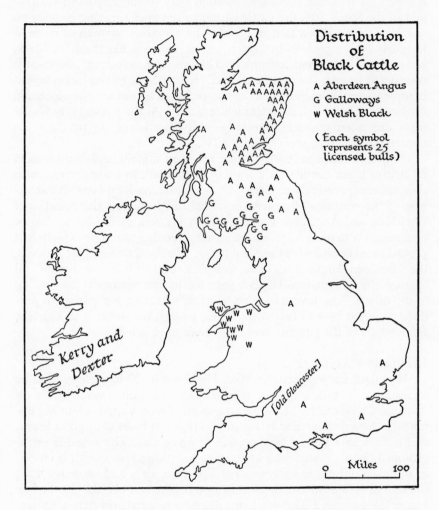

FIG. 25
The Black Cattle of the British Isles and their distribution

the other breeds of black cattle already considered the Galloway breed is noted for its great hardiness—second only to the Highland cattle—and in the home area the cattle winter in the open, grazing happily on rough pasture. Protection is afforded by a marked growth of more or less curly hair, especially in winter, and the breed is, like the other black, believed to be of great antiquity. Though slow maturing, the meat is considered among the best of all beef. Evidence of mixed blood is seen in some variety in colour, but the two pedigree strains are the black and the belted (black with a central white band). At the present day Galloways are often crossed, especially with Shorthorns, to produce the famous polled "Bluegrays" (Plate XII).

Although the *Aberdeen-Angus* are the most famous and important of all British black cattle and a leading beef breed, in their present form they are comparatively modern. Black cattle have long been characteristic of the counties of eastern Scotland, in particular the counties of Aberdeen and Angus (or Forfar). Hugh Watson, of Keillor in Angus, established a herd in 1808 adopting the breeding methods which had proved so successful with the Shorthorns. Unlike the other black breeds, the Aberdeen-Angus is hornless. Another contrast is in the rearing and feeding : the Aberdeen-Angus is yard fed in contrast to the hill feeding of the others. The favoured type is small, maturing early so as to produce excellent beef at two years. Thus, though the strain is an old one, refinements of the present breed are modern (Plate 6).

HIGHLAND CATTLE (Plate 5)

Highland cattle, properly West Highland or Kyloes, deserve to be mentioned at this stage as representing the other main strain of aboriginal wild cattle. The animals though of most varied colour are unmistakable with their sturdy low-set body, short head with great horns, and a thick skin with a long growth of wavy hair. The breed is extraordinarily hardy. American experience has shown how the air occluded in the mass of long hair prevents chill of rain, sleet and snow reaching the skin as it does sooner or later in nearly all other breeds. In the late winter days of the Dakotas when sub-zero temperatures may alternate with figures up to 60° or 70° a proportion of most cattle succumb ; not so the Highland. Though seemingly large they are actually the smallest of the beef breeds and mature slowly. But though they pay for finishing on good pasture, they will produce excellent beef at four years old without any subsidiary feeding—just what they find on the hill pastures. In the open they are nervous rather than fierce animals. Where they live natural lives—the bulls running with the cows and calves—on the American ranges they become extraordinarily tame. In Britain efforts to improve the breed by the introduction of alien blood failed and this points to their derivation from a wild ancestor, believed to be *Bos primigenius*, different from the ancestor of other British breeds.

WHITE CATTLE

Britain has no breed of white cattle if we except "park cattle" including the famed Chillingham herd of Northumberland, retained in a semi-wild state for centuries. The old white cattle and their modern descendants have recently been studied exhaustively by Whitehead (1953), though Lord Tankerville, President of the Breed Society, does not agree with all his findings.

THE RED BREEDS

The red breeds—the Sussex, South Devon (or South Hams), North Devon, Hereford and Lincoln Red—may all have been derived from a common stock, the red cattle brought by the Anglo-Saxons. Perhaps the *Sussex*, still of considerable importance in Kent, Surrey and Sussex, is nearest to the original, because it is a hardy and sturdy animal which one can well picture yoked to the ploughs with which the settlers tried to tame the intractable Wealden clays. Indeed it was bred for draught purposes until the early part of the 19th century. My own grandmother remembered Sussex oxen ploughing (shod with iron shoes) on the farms at Heathfield in her girlhood about 1840. The Sussex were probably the last cattle regularly so used in England,* though oxen were regularly used on Fair Isle till the Second World War.

Closely allied to the Sussex is the *Devon* or North Devon, a sturdy, red, dual-purpose animal though actually rather smaller than the Sussex. The lighter red *South Devon*, native to the southern half of the county, is the largest of the breeds of British cattle. Cows often weigh 15 or 16 cwt. (1,680-1792 lbs.) and bulls 30 cwt (3,360 lbs.). The rich milk is in part responsible for the fame of Devonshire clotted cream and the breed may owe its milk yields as well as its pale red colour to some admixture with Channel Island breeds (Plate 8).

In many ways the *Hereford* is the most famous and one of the most distinctive of all British farm animals. Herefords have long been an export of Britain much appreciated in both North and South America,

*A team existed at Birling Manor, Eastdean, Sussex, till 1929 and is illustrated by M. E. Seebohm (1952). Earl Bathurst had a team of Herefords near Cirencester till later.

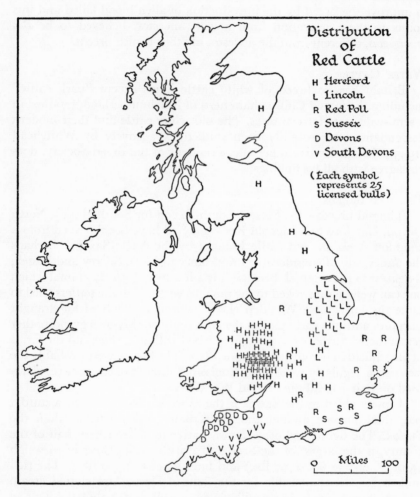

FIG. 26
The Red Cattle of the British Isles and their distribution

where they have proved themselves the ideal range animals. Remembering the Herefords at home on the lush pastures of their native county, it is curious that their American reputation should rest so largely on the ability to withstand shortages of both water and fodder. Actually at home they are bred under very natural conditions—running on grass in the open air throughout the year and receiving only a little extra hay or straw during severe weather or at calving time. At home too, as on the range, Hereford cows make excellent mothers. Though derived from common stock with the Devon and Sussex, the characteristic white face (which is very persistent when Herefords are crossed) originated from a cross with white-faced Flemish cattle imported into Britain in the 17th century. The breed as known today was stabilized by two breeders, Benjamin Tomkins, father and son, between 1742 and 1815. They were developed as beef animals to meet the new demands for meat from the growing industrial towns of the time and beef animals they have remained (Plate 1).

The Lincoln Red Shorthorn stands in rather a different position from the other red breeds. It is in fact a strain of Shorthorn selected for its cherry red colour and bred in Lincolnshire for the past century.

THE LONGHORNS

The Longhorns were once the dominant cattle of much of Britain, especially of the northern English counties, as Markham recorded in Elizabethan times. The failure of Bakewell with the slow-maturing Lincoln-Longhorn really marked the beginning of the end for the Longhorns and so of what may have been a strain of domestic cattle going back to *Bos primigenius*. The Highland strain, derived from the same ancestor, of course persists.

THE POLL BREEDS

The Poll breeds—in other words hornless cattle—are of mixed origin. It should be noted that any of our breeds of domestic cattle can be, and now usually are, dehorned or the horns of calves prevented from developing, but there are certain breeds in which the hornlessness is hereditary. The Aberdeen-Angus has been noted as outstanding but of the polled cattle, believed to have been introduced by the Scandinavians in pre-Norman times, the chief survivors today are the Red Polls of East Anglia: the only hornless breed of English as distinct from Scottish cattle. The original polled animal was the Suffolk dun, noted for its good milking properties, and it was not till the 19th century that this was crossed with the red horned Norfolk breed to produce the Red Poll, first recognized as a distinct breed in 1846.

THE SHORTHORNS

The Shorthorns may perhaps claim to be the real "Britons" amongst cattle. In the veins of most men of this realm flows blood which mingles Celtic and Anglo-Saxon, perhaps Roman and Scandinavian as well as Norman and other strains. So it would seem the brothers Colling built the Shorthorn breed by selecting the best of cattle then in existence in the country. Breeders who followed the Collings placed the same reliance on selecting bulls of quality proved by the characters of their off-spring and using them for a series of years in building up a herd. A mixed ancestry for the Shorthorns is suggested by the wide colour range —reds, whites, roans and many and varied mixtures. On the whole the Scottish Shorthorns are beef animals, the English dairy animals, but the separation of the beef and dairy breeds is impossible.

THE DAIRY BREEDS

A major part of our milk supply is now derived from British Friesians and a certain quantity from dual-purpose breeds, especially dairy Shorthorns. But there remain three great dairy breeds—the Ayrshire, Guernsey and Jersey. Of these only one, the *Ayrshire*, is a native British breed. Like the Shorthorn, the Ayrshire is of mixed parentage. It probably includes northern Polled cattle and early Dutch imports as well as some Highland and perhaps Channel Island blood. It was gradually stabilized by the gentlemen farmers of Ayrshire during the 19th century till the modern animal with its upward inclining horns and brown and white body colour became very distinctive. Though some animals may show black patches, in general the Ayrshire may be described as white with very varied patches of reddish brown varying from a few near the head to many covering most of the body. It is a hardy animal and, after 1919, increased in popularity in England and Wales, well away from its home area till supplanted by the Friesians. There is a good milk yield with a steady average of 4 per cent. butter fat. The fat globules are very small so that the cream does not readily separate from the milk but the latter is often as a result pale in colour. The finely divided fat makes the milk especially suitable for infants (Plate 7b). Animals are now commonly dehorned.

The *Jersey* is a small, graceful animal of varying colour—fawn, brown and grey—but commonly and distinctively fawn with a darker brown around the head. The milk yield is only moderate in quantity, extraordinarily rich, the average of butter fat being well over five per cent. As a result the milk is a deep rich colour and the cream separates easily. Jersey milk is often sold separately at a higher price than other milk but may be so rich as to be indigestible especially by infants. Though possibly originating from the neighbouring parts of the French mainland, the breed has been perfected in the small island of Jersey. It affords a perfect example of controlled breeding by careful selection. The island, intensively cultivated and densely peopled, is less than 40,000 acres and for nearly 200 years (since 1763) the importation of live cattle into the island has been entirely prohibited, so that all the cattle are pure bred. There is of course a steady export of live cattle and Jersey herds are found in many parts of southern England. A combination of damp and cold is disliked by Jerseys but otherwise they are hardy and, surprisingly enough, the bulls have a reputation for ferocity.

The *Guernsey* is a larger animal than the Jersey but the breed has been kept pure by similar means in the Channel Islands of Guernsey, Alderney and Sark. Guernsey milk is also very rich and of a deep colour Most Guernseys are fawn in colour, characteristically lighter than the Jersey, usually with some white.

The fame of Dutch cattle as milk producers goes back to the Middle Ages and there were doubtless spasmodic and irregular importations of animals from Holland over several centuries resulting in an infusion of Dutch blood into many British breeds. Rather naturally the influence was strongest in the east and in the early part of the 19th century the Holderness cattle included many black and white animals resembling the modern Friesian. In 1880 the importation of any cattle into the country for breeding purposes was prohibited. British breeders set to work to develop a distinctive dairy strain from Friesians previously imported and in 1909 the British Friesian Society was formed. In 1914 the importation of 50 specially selected animals from Holland was permitted and further importations came later from Holland and also from Canada and South Africa. As a result the British Friesian holds many records and has increased enormously in popularity. Cows have yielded 3,000 gallons of milk a year ; the large calves may weigh 100 lbs. at birth—so that males are soon of much value as veal. There is no mistaking the black and white animals—a background of white with irregular jet black markings. Some years ago an enterprising company published a picture of a Friesian with markings resembling a map of the land masses of the world, and the Friesian was dubbed "the world's cow" (Plate XIII).

SHEEP

No area of comparable size anywhere in the world can rival Britain in the number of breeds of sheep. All those noted in the accompanying table are recognized breeds and nearly all have their appropriate breed-societies. In addition there are local types not included in the list. The importance in recent years of the problem of hill-sheep farming and the adequate use of marginal land has led to exact figures being available of the principal hill-breeds. Of lowland breeds no census exists but through the breed societies the relative importance is known as well as the distribution of the chief flocks. Some breeds have become of small importance (such as the Limestone and the Penistone) and may become extinct. In addition to the established breeds there is much deliberate cross-breeding especially of Cheviot and Leicester, which gives the Cross-Bred sheep (spelled thus with capitals) as well as casual crossing.

The domestication of the sheep dates from very early times and is believed to have taken place in the general region of south-western Asia or the borders of India. There are numerous wild species of the genus *Ovis* or true sheep and they are or were widely distributed in both the Old and New Worlds. Two groups are commonly distinguished. The first group comprises the mouflon (*Ovis musimon*) of Corsica and Sardinia, and a species of sheep with over a dozen races from northern India and Turkestan to Asia Minor and Cyprus that are now united under the name *O. orientalis*, the urial (Ellerman and Morrison-Scott, 1951). The second group comprises the argali (*O. ammon*) of central Asia, with which Marco Polo's sheep (*O. poli*) is now united and the bighorn (*O. canadensis*) of Siberia and the Rocky Mountains of North America. Carl Sauer in his recent review of the evidence regards the *vignei* group of urials of Persia, northern India and Tibet as the chief parents of our modern domestic sheep ; older views favoured the mouflon. The Merino may have been derived from quite different stock. In any case domestication was very early and little can be known for certain of the original stock from which the domestic sheep were derived.

In Britain Neolithic man had flocks of small, slender-legged long-tailed horned sheep—together with the oxen noted above, goats and pigs. The question arises what part did sheep play in this early farming economy ? It is almost certainly not as a source of meat—that was not to come for many centuries. Was it primarily as a source of wool or was it, as some believe, for the sake of the ewe's milk ? There is little

BREEDS OF SHEEP
(after J. H. H. Thomas and others, *Sheep*, London. Faber & Faber, 1945)

GROUP A. *Mountain and Hill Breeds*

c Scottish Blackface	Scottish Highlands etc.
Swaledale	⎫
Rough Fell	⎪
c Lonk	⎬ Breeds of the Pennines
Derbyshire Gritstone	⎪
Limestone	⎪
Penistone	⎭
Herdwick	Lake District
c Cheviot	Southern Uplands etc.
Welsh Mountain	Wales
Exmoor Horn	Exmoor
Shetland	Shetland

GROUP B. *Intermediate Breeds linking A and D*

Kerry Hill	⎫ Breeds of the English-Welsh
Clun Forest	⎬ Border
Radnor Forest	⎭

GROUP C. *The Short-Woolled Breeds*

Dorset Horn	⎫ White-faced sheep
Wiltshire Horned or Western	⎭ with horns
Ryeland	⎫ White-faced sheep
Devon Closewool	⎭ without horns
Southdown	
Shropshire	
Suffolk	⎬ Dark-faced sheep without
Dorset Down	horns—the "Down" breeds
Hampshire Down	
Oxford Down	⎭

The last-named is larger and longer in the fleece and hence forms a link with

GROUP D. *The Long-Woolled Breeds*

Leicester	Yorkshire Wolds
Border Leicester	Midlands
Lincoln Longwool	Lincoln Wolds
c Wensleydale	Yorkshire Dales
Cotswold	Almost extinct
Kent or Romney Marsh	Kent
c Devon Longwool	North Devon and Cornwall
South Devon	South Devon and Cornwall
Improved Dartmoor	Dartmoor

NOTE.—At the 1963 Royal Show some of the mountain breeds were, naturally, not represented; Wensleydale, Cotswold and Devon Longwool were cancelled classes; Teeswater were added.

evidence that ewe's milk was the primary product: interest centred rather in the wool from the live animal, and the skin (for clothing, later for parchment) and the fat for tallow from the animal when dead. Natural sheep skins are extraordinarily warm as well as light and pliable and perhaps better for clothing than the skins of any wild animals available in Britain.

The Neolithic sheep moulted naturally in the summer ; they were not shorn. The Soay sheep of St. Kilda is, in a sense, a living fossil of the Neolithic sheep. The animals were not shorn, the wool was plucked. Soay flocks have now been introduced on to the islets of Ailsa Craig and Skokholm.

Wool when scoured, that is with the natural grease washed out, felts or mats and so forms a fabric which may have been in use even before spinning and weaving were either known or became general. The fescue pastures of those well drained downlands on which Neolithic and early Celtic farmers made their homes are still ideal sheep pastures and it is likely that the animal soon gained favour. There may even have been an export of wool from Britain in pre-Roman times and the stage was set for a trade destined to last, with interruptions, for a millenium and a half.

Evidence points also to an early differentiation which may have taken place in Britain of the three strains of sheep—the hill sheep, suited to mountain pastures, the short-woolled sheep of the downs and the heavier long-woolled sheep associated with the rich cornlands of the lowlands and East Anglia.

This long list of recognized British breeds may confuse and frighten the layman but fortunately, as well as the fairly obvious division into four groups, most breeds of sheep have distinctive faces. (See Plates XIV and XV.)

Footnote

The terminology used by sheep-farmers is very confusing to the layman and, further, differs widely from one part of the country to another. The young animal up to about nine months is a lamb, though by that age some breeds are so large that the meat would be sold as mutton. Female lambs according to the district are referred to as ewe lambs, gimmer lambs or chilver lambs. Male lambs are ram lambs or tup lambs but after castration (which is usual except for the rams for breeding) they become wedder or wether lambs. Most sheep are shorn for the first time in the second summer of their lives (some southern breeds in the first summer) and between the end of lambhood and the first shearing are known as hoggs, hoggets or tegs. As a hogg may be a ewe, a ram or a wether, nomenclature at this stage is complex, more especially as such synonyms as a "diamond tup" may be used for a teg ram.

A shearling is an animal awaiting its first shearing at 15 to 18 months, but young shearling ewes in the north are called gimmers or, in parts of the south, theaves.

A ewe produces her first offspring at two years of age : a ewe that has lambed for three seasons is middle-aged, one that has produced lambs for six seasons is distinctly old. A ten-year old ewe is senile : to reach 12 or 13 years is equivalent to being a venerated nonagenarian.

The period of gestation is 147 days or 21 weeks and a crucial period is the beginning of the breeding season when the shepherd "makes up" his flock. For January lambs in the south July is the crucial month. It is then that any ewes which show defects, especially in udder or incisor teeth (which would impair their ability to feed and keep in conditions during the winter), are taken out or "culled", In breeding it is usual to allow one ram to 50 or 60 ewes. Ewes are equipped by nature to suckle twins, but breeds vary considerably in the proportion of twins : 140 lambs per 100 ewes would be good for most breeds.

Mountain and Hill Breeds

The various breeds of hill sheep have perhaps little in common except their generally small size and their ability to survive and even thrive on mountain moorland grazing. The first four in the above table —Scottish Blackface, Swaledale, Rough Fell and Lonk—all have black faces, are horned and have coarse wool. The Derbyshire Gritstone has softer wool and is hornless. All the hill breeds yield good meat—small joints of lean, sweet meat—and occupy a special place in British farming because they permit an economic use of hill lands of otherwise restricted usefulness and provide healthy breeding stock for crossing with lowland breeds.

Surprising though it may seem the *Scottish* or *Scotch Blackface* is a comparatively recent immigrant into the Highlands of Scotland. Claims have been made for Ettrick Forest as the original home of the breed and it has long been at home in the Southern Uplands, but was not introduced into the Highlands until about 1770. That was the time when the Highland grazings were turned over from mainly cattle (with a few of the old Celtic soft-woolled sheep now found only in the Western Isles) to form great sheep runs. The Blackface are active, even wild, and roam widely, grazing especially on the heather-clad "black" hills (as contrasted with the "green" hills which are grass moorland) up to 3,000 feet. At higher levels they are kept pure : various crosses are favoured at lower elevations. The ewes live out on the hills throughout the year and get no subsidiary feeding except a little hay when the ground is covered with hard snow. The higher ill-drained levels of Scottish moorland are largely clothed with cotton-grass, *Eriophorum,* and if the sheep eat too freely the lambs develop milk sickness from which

mortality can be very high. The delightful remedy quoted in the old
books is a teaspoonful of good Scotch whisky twice daily. The face and
legs of the Blackface are not always black. They may be "brockit"—
with white patches. In 1941 Blackface accounted for 1,703,000 of the
2,270,000 hill sheep in Scotland, or 75 per cent. (See Plate 9.)

The *Swaledale*, found in Swaledale and the neighbouring Pennines
westwards into Westmorland, is rather larger than the Scottish Blackface and the nose is usually grey, the legs spotted.

On more exposed situations especially in an area centred on Kendal
the Swaledale and other sheep give place to the extremely hardy *Rough
Fell* or *Kendal Rough Sheep*, larger than the Scottish Blackface, shorter
woolled, and usually with a brownish tinge on the face.

It is a usual custom of many hill sheep farmers to leave the sheeps'
tails undocked. The tail of the Blackface is comparatively short whereas
that of the *Lonk* reaches almost to the ground. This breed, of the central
part of the Pennines, often with much white on the face, has a finer,
denser and heavier fleece than the Blackface, but it is not nearly so
hardy. There is some evidence that the type of fleece is a response to
particular aspects of the climatic environment and that of the Lonk
would seem to be rain-resistant rather than a protection against snow.

The rough carpet wool of the Blackface comes from a fleece which
averages about 4 lbs. : the wool of the Lonk goes for blanket manufacture.

The *Derbyshire Gritstone* from the Millstone Grit lands of the north-
west of the county is a transitional type—larger than the Blackface,
hornless, the face and legs black with white markings, the fleece finer
and softer.

The *Limestone* and *Penistone*, once important, have now become
virtually extinct in the face of Blackface competition.

Those who love the fells of the Lake District know the *Herdwick* as the
most characteristic animal. The lambs are born with black faces which
later turn grey and become white in the fully adult animal. In a land of
driving rainstorms and scanty rough grazing the Herdwick has no rival
for endurance. The wool is very long but coarse and often streaked with
grey. Very little is known of the origin of this old distinctive breed:
tradition ascribes its advent in Cumberland to the wreck of a Spanish vessel
on the coast but the Rev. T. Ellwood suggested a Scandinavian origin.

Whether or not other sheep acknowledge the *Cheviots* as the aristo-
crats of the sheep world it is impossible to say, but they always contrive
to look that way. The clean white face, with its ruff of wool, erect ears,

the Roman-nose effect and a slightly supercilious expression, together with a proud carriage of the head all combine to impress the most casual observer that here is a breed of great importance. A few males are horned, but the ewes are hornless. It is primarily the sheep of the green (i.e. grass) hill pastures of the Cheviots where it has existed from very early times and may be descended from the aboriginal tan-faced Soay which some have called the Celtic sheep (*Ovis celticus*). From this centre the Cheviots have invaded lands both north and south, from Sutherland and Caithness in the north, to Wales, Exmoor and the Chalk downlands in the south. Those which have migrated to the north have become somewhat different and now have a separate flock-book. Cheviots require better conditions than the Blackface and are found on the lower, drier and more grassy hills, and they need subsidiary feeding in the winter. On the better hill farms they are crossed with lowland breeds, especially Border Leicesters.

The hardy *Welsh Mountain* sheep is also of pure Celtic stock and is at home all over the Welsh mountains. The very active, rather wild animals love their mountains and do not take kindly to restrictions. The rams are horned, the ewes hornless and in both the white faces often have tan patches. A distinction may be drawn between the true mountain animals, brought down to lower levels for wintering and lambing and the rather larger animals with a short, fine dense wool kept on lower ground. The true mountain sheep have a fleece of only about 2½ lb. which is often coarse but it is claimed that the mutton is of incomparable flavour and quality. (See Plate 10b.)

The ancient breed of the *Exmoor Horn* is native to Exmoor and the Brendon Hills. The animals are small, horned in both sexes, with white face and legs but a black muzzle.

The *Shetland* affords an interesting example of localization being restricted to the Shetland Islands and even there it has a rival in the Blackface. It is small, slow maturing and varied in colour and the fleece is of varied quality. Part of the fleece however is of extremely fine soft wool which is picked off at intervals when it becomes loose instead of the sheep being shorn. In this way the fine wool is obtained for hand-spinning and then the far-famed local industry of shawl making. It is possible that the Shetland, like the Soay sheep of the St. Kilda group, is a direct descendant of the Neolithic animals.

THE INTERMEDIATE BREEDS

It is interesting to note that in the transition lands between Highland

and Lowland Britain, on the borders of mid-Wales, three separate breeds of sheep have been evolved intermediate between the short-woolled mountain breeds and the long-woolled lowland breeds. The Kerry Hill, Clun Forest, and Radnor Forest each takes its name from the small home area. Especially amongst the Kerry Hill there is local variation apparently according to habitat conditions.

THE SHORT WOOLLED BREEDS

These are the sheep of Lowland Britain and are associated with arable or mixed farming. The *Dorset Horn* is an ancient breed with finely developed curved horns in both sexes and easily distinguished by the flesh coloured lips and nostrils. An average wool clip from ewes is 5 lb. —of a clean, white wool. Before the days of the importation of New Zealand and Australian lamb the Dorset Horn was particularly valued because the ewes could be mated in April or May to produce lambs in October in time for Christmas fattening. Despite the forbidding horns the sheep are docile and the ewes are both prolific (150 per cent. of lambs) and make good mothers.

The *Western* or *Wiltshire Horned Sheep* is another old breed which almost became extinct but has been revived—not in its native county but in Northamptonshire and Anglesey as small flocks. There is no mistaking this breed for it produces practically no wool, and what little there is falls off in spring. (See Plate 10a.)

The old breed, the *Ryeland*, takes its name from the rather poor light-soiled lands, the ryelands, of southern Herefordshire. Ryeland sheep are small animals with a very fine wool but nowhere are they now of first importance. By way of contrast the *Devon Closewool* is a newcomer evolved in the present century (Breed Society 1923) by crossing the Devon Longwool and the Exmoor Horn. There is an extensive area of land in north Devon which though below the moorland level can only provide indifferent grassland grazing : the Devon Closewool was the breeders' answer to these conditions.

The six Down breeds have much in common : they are hornless and dark-faced and have been evolved from the shortwoolled sheep which from time immemorial had roamed the chalk downlands of southern England or had eked out a precarious existence on the aftermath or the stubble of the common arable fields. Pride of place must be given to the *Southdown* which John Ellman of Glynde, near Lewes, worked on for fifty years to produce from unpromising material a breed unsurpassed the world over in quality of meat combined with rapid maturing and a

Plate 7a. WELSH BLACK CATTLE. By the River Conway (*p. 130*)

b. TWO PEDIGREE AYRSHIRE HEIFERS. A dairy breed of considerable popularity; the animals are now commonly dehorned (*pp. 121 and 136*)

Plate 8. RED DEVON CATTLE

very fine 2½ inch wool, which commands the top price amongst British wools, from a 4½ lb. average fleece. Ellman started farming about 1780 and used breeding methods rather different from those of Bakewell in that he used less close inbreeding and concerned himself with both wool and meat. Ellman's work was carried on by Jonas Webb of Babraham, Cambridgeshire, and others in the earlier part of the 19th century. The Southdown is quite a small animal even compared with other Down breeds but is neat and compact in build with a broad, even, mouse-coloured face with wool on the forehead.

The *Shropshire* was evolved in that county by crossing of local sheep with the improved Southdown and was recognized as a distinct breed, since stabilized, in 1859. It is intermediate in size between the small Southdown and the larger Down breeds and has a soft black face and legs. The dense fleece averages 7 to 8 lb. of fine quality wool of 3 inches or over. The mutton is good but tends to run to fat.

The *Suffolk* likewise originated from a Southdown cross—in this case with the old black-faced Norfolk Horned sheep now extinct. Though the breed was recognized locally by 1850, separate classes at the Royal Show were not established till 1886.

The *Hampshire Down* is another Southdown cross—with the old dark faced horned Berkshires, now extinct, and the Wiltshire. It was first recognized in 1861. The face and legs are a deep rich brownish black. The animal has shown itself particularly adapted to folding on the light arable lands of southern England. The *Dorset Down* has a lighter face and the breed was evolved by mating Southdown rams and Hampshire ewes and then crossing the lambs produced with an old native western Down type.

The *Oxford* is the largest and heaviest of the Down breeds and is the only case where a cross between the Down breeds and Longwools has been fixed. The breed was recognized as early as 1862 and is still characteristic of the Cotswolds, where it replaced the original Cotswold Longwool. The fleece is heavy—9 lb.—but the meat tends to become coarse and poor after the animals are a year old.

THE LONG-WOOLLED BREEDS

The decline in popularity of the traditional British lowland sheep over the years of the present century is an interesting commentary on changing tastes. They produce fleeces of 10 to 15 lb. in weight the wool being lustrous and long-stapled, though large in fibre diameter. Such wool now commands smaller prices than the finer, shorter-stapled wools.

M.L.—L

The long-woolled sheep produce rather large joints of meat tending to coarseness of grain and fatness. The days have passed when the house-wife had a large family to feed and melted down the mutton fat for cooking over the rest of the week. When that great pioneer, Bakewell, evolved the improved Leicester in 1755 to 1790, it was to meet the meat needs of the growing industrial towns : he could afford to sacrifice wool for large fat joints. The modern Leicester now most popular on the Yorkshire wolds is still a good animal—ewes clip about 10 lb. and the meat is of good average quality. Nevertheless the Leicester today is chiefly valued for crossing.

From about 1800 onwards Bakewell's Dishly Leicesters were sent to Scottish border districts where the *Border Leicester* was evolved. It is not clear whether there was crossing with the Cheviots : if not the Leicesters learnt effectively to ape the Cheviots by developing a Roman nose and more erect ears, and losing the wool on the forehead. The famous Half-Bred (spelt thus with capitals) which are the very popular products of Border Leicester rams and Cheviot ewes have the Roman nose and clean white face with erect ears extraordinarily accentuated.

The *Lincoln Longwool* is the largest British sheep and mature rams may reach 400 lb. The long wool may even reach 18 inches in length and a good flock will give an average weight of 12 lb. in fleeces. Ringlets of wool hanging over the face are a distinctive feature. Except from young lambs the meat however is poor. Lincoln farmers on the Wolds and Heaths remained faithful to the breed but it was little seen else-where in Britain till the Second World War, when flat-rate prices per pound for all sheep focussed farmers' attention on the Lincoln. But the Lincoln has been of immense value to Australia, New Zealand, and the Argentine for crossing with Merino and is claimed as the basis of the famous Corriedale Breed of New Zealand though there is some doubt about this.

The *Wensleydale* resulted from crossing improved Leicesters from Dishly with an old local breed and the type was fixed about 1860. It is a distinctive animal with a bluish-grey face and a very silky fleece which hangs in delicate pencil-sized ringlets. It is able to withstand harder conditions than its Lowland parent and so is a long-woolled breed adapted for higher poorer grazings such as are found in the Yorkshire Dales area.

Alas, the famous old *Cotswold* is now almost extinct, but not so that remarkably fine animal the *Kent* or *Romney Marsh*, now the chief from the point of view of numbers of our long-woolled breeds. It was evolved on that remarkable stretch of country on the borders of Kent and Sussex known as Romney Marsh. There one finds a curious combination of bleak windswept land, especially in the winter, with some of the finest grasslands to be found anywhere in the country. So these short, compact, short-legged but heavy sheep with a fleece of some 8 to 9 lb. of wool which is finer in quality, shorter and denser than any of the other Longwools, flourish under conditions of rich but exposed grassland. Large numbers are sent to neighbouring upland farms for wintering but they can and do fatten entirely on grass and the lambs particularly yield excellent meat. In Britain the breed is restricted to the south-east but is important in nearly all the great sheep lands of the southern hemisphere—South America, including the windswept Falklands, New Zealand and Australia.

The *Devon Longwool* is a localized breed of the better lands in the south-western counties ; the larger *South Devon* takes its place in Cornwall and South Devon. The *Improved Dartmoor*, though evolved from a moorland breed, has really become a Longwool largely with the help of Leicester blood. An interesting question is why these three breeds should all be localized in the south-west. It would seem they agree in being able to flourish in the humid but mild climatic conditions of the regions, each in turn adapted to slight differences in physical conditions.

The portrait gallery (Plates XIV and XV as well as Plates 8 to 11) will enable most of these British breeds of sheep to be identified but the great prevalance of cross-breeding must be stressed—usually to emphasize one or more characteristics. For controlled cross-breeding to be successfully continued, pure breeding stock must still be maintained. Often the reason for cross-breeding is to secure good quality, quick-maturing lambs for the butcher. Cross-bred animals often develop what has been called 'hybrid vigour'—a greater resilience than either parent to environmental conditions. Broadly speaking to mate a cross-bred with a cross-bred does not produce standard results, so that there must be continued recourse to pure parents. The famous Greyface or Mule produced by a Border Leicester with a Scottish Blackface ewe is however a good breeding animal. Most sheep experts agree with controlled cross-breeding but deplore the existence of mongrel flocks.

PIGS

According to Sauer's recent review of the evidence, "the wild pig of the southeast Asiatic Mainland, *Sus vittatus,* is unquestioned as ancestor of the domestic pig of that part of the world . . . . . (it) lives in jungle forests, likes to root in the village plantings, and does not avoid human settlements as does the European wild-boar."* Like the dog, he continues, "the domestic pig too has become a member of the household and its domestication may have been by the same route as that of the dog." If one draws a distinction between herd animals—cattle, sheep and goats—and the more strictly domestic animals associated with *domus*, the home itself, then the pig joins the dog, cat and poultry in the family circle. Certainly Neolithic man in Britain had the domestic pig, destined to increase steadily in importance and quite distinct from the savage and dangerous wild boar of the woods, destined to extinction. This position is paralleled by that of the domestic dog and the wolf at the same time. In both cases there may have been an occasional admixture of blood.

Whilst for many centuries cattle were valued as draught animals and for their hides, sheep primarily for their wool, it was the flesh of the pig that rendered the animal valuable to man. Catholicity in food habits, fecundity in breeding, rapidity in maturing, richness in fat, ease of meat-storage by salting and smoking have all favoured the use of the pig as a source of human food from early times. It is curious that the rise of the concept of the pig as an unclean animal has led to the elimination of such an easy source of food throughout Hindu, Jewish and Islamic communities. This may well have originated, not so much because of the omnivorous feeding habits of the pig as the rapidity with which the fresh meat deteriorates, especially in hot climates and becomes a dangerous source of disease. Until quite recent times the British housewife would not buy pork unless there were an 'r' in the month—thus eliminating the summer months May to August inclusive.

Pork was extensively used by the Romans and roast pork was the main meat dish at Roman banquets. For many centuries the British appreciated also the flesh of the wild boar which was hunted until the latter part of the sixteenth century.† The boar's head became symbolical both of the joys and fruits of the chase and of the festivities which followed the hunt. It survives as a favourite public-house name.

The majority of townsmen born and bred are able to recognize a Hereford or a Jersey but a pig is just a pig. The dozen breeds now com-

---

*Ellerman and Morrison-Scott (1951) however consider the Asiatic and European wild pigs as two races of *Sus scrofa*.

†A wild boar was killed in Staffordshire in 1593 and Harting (1880) gives evidence of survival in northern England until the reign of Charles II (1660-1685).

Farmer & Stock-Breeder Photograph

Plate XVII.—A Ploughing Match at London Colney, 1938

It is frequently no longer possible to hold such ploughing matches because of the almost universal adoption of tractor ploughing (*p. 153*)

L. Dudley Stamp

Plate XVIIIa.—New Forest Ponies. Grazing on a clearing in the Forest (*p. 154*)
b.—Harvesting above Chideoak, near Bridport
This photograph was taken in 1938. The use of farm horses and a four-wheeled farm
cart of this type is fast becoming rare

The Times

mon in Britain are known mainly to farmers. Curiously enough it would seem that for a couple of thousand years a pig was just a pig even to farmers, and little effort was made to improve the domestic pig. There is evidence of domestication in Neolithic times and the pig was held in high esteem by the Anglo-Saxons. Under Manorial England every village had its swineherd and one of the Domesday methods of recording woodland is in terms of the 'pannage' afforded to pigs—that is pasturage where pigs lived on the acorns of the oak trees or the mast of the beeches. Such representatives of the pig as have come down to us show a long-legged razor-backed animal with long hair—not very different from the wild boar. Some improvement must have taken place : during Tudor times sties were built to house the pigs though they were turned out by day on to the waste lands and into the woods. By the early part of the eighteenth century pigs of a large white type existed. These supplied the material for Bakewell's breeding experiments which began in 1760. It was shortly after that, between 1770 and 1780, that an event of great significance took place. It was the importation of pigs for breeding from China and Italy. From very early times pigs had been bred in China—short-legged, fleshy animals, very different from the old British animals. These Chinese pigs had reached the Mediterranean area in Roman times and had been used in the development of the Neapolitan and other breeds. They were probably descended from the wild pig races of south-east Asia (*Sus scrofa vittatus*), whereas the indigenous pigs of Britain had been derived from *Sus scrofa scrofa,* the European wild pig.

It may be said that Bakewell changed the large white pig into the Large White, known also as Yorkshire, and so established the classification of individual breeds. It was not, however, for nearly a century that pig-shows used a breed classification : it was in 1851 that Joseph Tuley, a weaver of Keighley in Yorkshire, exhibited 'large white' pigs at the Royal Show and established the fame of the breed. Volume I of the Large White Herd Book was published in 1881 and in it the frequent mention of Tuley's boar, Sampson, shows—as with cattle—the value of the proven sire. Today the Large White is numerically the main British breed, and large numbers of breeding animals are exported, for the breed is very popular in Europe, Australia and elsewhere. It furnishes excellent quality lean bacon. The white skin gives the carcases a clean appearance. The docile sows produce an average of ten or eleven pigs at a litter, out of which eight are successfully reared, for the Large White make excellent mothers. The pigs gain between 0.8 and 1 lb. a day (Show animals as much as 1.3 lb.) and are ready for killing as pork from 145 to 160 days when the animals weigh 120 to 145 lb. (75 to 80 per cent. of which is usable meat.) In 200 to 230 days they have reached 200 to 230 lb. and are ready for the bacon curer. (See Plate XVIb.)

The *Middle White,* a pork type pig for small choice joints, was also developed by Joseph Tuley, who crossed a boar of the former Small White with sows of Large White breed.

The *Berkshire* also was established as a breed at an early date, the first Herd Book being published in 1885 ; its development took place near Wantage.

The *Wessex Saddleback* (Plate XVIa) is a distinctive animal which originated in the Isle of Purbeck and has spread over the southern counties from Devon and Somerset in the west to Sussex in the east. It is an ideal out-door pig. Curiously enough the first Herd Book was not issued until 1919. Similarly the first Herd Book of the allied *Essex* dates from 1919, though the breed, perhaps developed from the mediæval forest-feeding pig, had been known locally in Essex for at least a century.

Another old breed is the *Large Black,* popular in the south-west and in East Anglia. The Breed Society was established only in 1899—a century after the breed was clearly distinguishable. The white, drooping-eared *Welsh* pig (Society 1918) seems to be of the general type kept in Wales for centuries. Being an excellent bacon pig it has spread outside Wales in recent years.

The *Tamworth* is an interesting example of a very old pure breed (Herd Book 1885) though not very numerous. The colour is said to have been fixed by the importation of an animal from India about 1880 by Sir Francis Lawley of Middleton Hall, Tamworth.

Other pig breeds in Britain are mainly of local importance, but often interesting because of their localisation and distinctive characters. The *Cumberland* (Herd Book 1916) was established at least by the early 19th century in the north-west and the border counties. The *Gloucester Old Spot* (Breed Society 1914) had also long been known in the west of England—especially in Gloucester, Wiltshire and Somerset—and achieved a meteoric rise to fame in the nineteen-twenties, only to fall quickly from favour. Similarly the *Large White Ulster* (Herd Book 1909) formerly universal in northern Ireland has been ousted by the Large White. The *Long White Lop* or Lop-eared of Devon and Cornwall (Herd Book 1920) has similarly given way to Large White or Large White crosses. The hardy, prolific and picturesque *Lincolnshire Curly Coat* has never been very popular outside its native county.

What are the lessons to be learned from this brief catalogue of British pig breeds? First, the localization of contrasted types resembles that of cattle and sheep. Second, it is difficult to suggest that environmental factors have been important in the differentiation and the localization unless one includes availablity of food of certain types as such a factor. Some pigs are essentially "outdoor", others "indoor". Third, the principle of up-breeding by the use of proven sires is as evident in pigs as in cattle and sheep. Fourth, different qualities are needed in a porker or pig to be eaten as pork and a baconer. Pig breeding is still in Britain something of a casual side line with most farmers. The Danes on the other hand have achieved their great success by careful standardization.

The number of pigs in Britain shows a remarkable fluctuation from year to year which has attracted the attention of economists and led to the now familiar concept of the pig cycle. It may be said that the pig has a certain number of disadvantages as a farm animal but many advantages as a domestic animal. Where feedstuffs are grown or imported the pig is a relatively poor converter in terms of protein yield of the meat, hence during World War II farmers were discouraged from keeping pigs. But pigs will thrive on household waste—swill from hotels and restaurants—and mature rapidly, hence the 300,000 small pig keepers of World War II. Pigs were kept on bombed sites in Plymouth and fed almost entirely on waste from the City's municipally-owned restaurants. Although the pigs did not share with the poultry the place of honour on the roof of the bombed City Hall, the City Engineer Mr. Paton-Watson developed a very effective system of collecting and processing waste food which should serve as a model of what is possible.

GOATS

Though goats have been domestic animals in Britain at least since Neolithic times, it is doubtful whether they have ever played a major role in the economy of the countryside since their Neolithic heyday. The domestic goat is probably derived at least in the main from *Capra hircus aegagrus*, the race of wild goat native of the mountainous regions of south-west Asia and the Greek islands, but the more easterly race *C. h. blythi* may have contributed. Those who draw a distinction between domestic animals strictly speaking and herd animals point out that the goat bridges the gulf. Whilst herds of goats are characteristic of dry Mediterranean lands where they find sustenance even on the poorest herbage and incidentally are extraordinarily destructive of any vegetation, goats in small numbers attached to individual households are far more widespread. Both pigs and poultry can be kept, for a major part of their food, on household scraps or kitchen waste—a fact which has rendered them of peculiar value through the ages and was not forgotten during the Second World War. In some respects goats are complementary feeders to pigs : they do not enjoy the "swill" of cooked food from the kitchen but will eat up the cabbage leaves and lettuces with rough hay of little use to pigs. The goat has sometimes been called the poor man's cow placing, rightly, the emphasis on milk. Whilst the early domestication of cattle was probably in the main if not entirely based on the need for a strong plough animal and the domestication of sheep primarily for their skins and wool, it would seem that the goat was valued from the start for its milk. In the natural state the wild goat is a timid denizen of mountain fastnesses. But the kids are easily tamed and become affectionate, if obstinate, household pets. This could clearly have happened in the beginning of domestication. Goat milk is rich and in ease of digestion by human infants is said to be superior to cow's milk and in certain aspects to approximate most nearly to human milk. So the goat has continued to hold its place as a household animal even to the present day—markedly so, indeed, in Britain during the Second World War. Unfortunately popularity is diminished by the strong smell associated with billy-goats and the somewhat strong flavour of goats' milk.

THE HORSE

Within the short space of the last fifty years the horse has lost the proud and dominant place occupied for two or three millennia in the life of man. Throughout that long time the horse was the personal servant of man rather than one of his farm animals. Indeed, the horse took the place of the ox as the draught animal for cart and plough only quite late in the history of British farming. As recently as the edition of 1834 the anonymous author of *British Husbandry*, a manual for farmers, discusses at length the relative values of horse and ox for the several needs of the farm and, with precise details of costing, comes down in general favour of the ox (pp. 185-6). But as a beast of burden carrying man himself, the horse reigned supreme till the advent of the bicycle, rubber tyres and the internal combustion engine. As the foundrous mediæval trackway gave place to metalled turnpike and to macadam, the horse became the undisputed draught animal for stage coach and goods wagon alike. The canals of the pre-railway canal era were scarcely a rival because there too traction demanded the horse. The railway killed the stage coach and long distance haulage of goods by horses, but not their use in local work. The development of electricity restricted the sphere of usefulness of the horse in several ways : electric trams for example, replaced horse trams of which some at least survived until the First World War ; later came the decline of the pit pony.

At the present day it has become almost a rarity to see a working horse in city streets ; the disappearance of the pit pony is depopulating equine Dartmoor. The decline of farm horses has been spectacular. The total of all horses, grazed on agricultural land in Great Britain dropped from 1,002,000 in 1938 to 494,000 in 1953. Out of these 668,000 were working horses in 1938, only 347,000 in 1953. Tiue it is urged in even such farm manuals as Fream's *Elements of Agriculture,* 13th edition (1949) that there is still an important place for the horse on the farm, but the defence reads rather like the defence of the ox in 1840. The argument why use a 20 h.p. motor to do work appropriate to a 1 h.p. horse ignores the truth that where there is a horse there can be no six-day week. So the farm horse would seem to be doomed and we are almost back where we started: a past when dignity and prestige demanded that the Roman officer should be equestrian and a present when the sport of kings promises to remain even more firmly than kings themselves. By 1963 there were only 100,000 horses on farms in Britain.

One species of wild horse, the tarpan (*Equus przevalskii*), became extinct in Europe only a few decades ago. Prior to that it roamed the south Russian steppes, and still exists wild in Mongolia. Extinct at an early date were the heavier horses once wild in the wooded lands of north-western Europe. Crossing of the two stocks is believed to have produced the heavy carthorse.

It is certain that the early horses of Lowland Britain in Roman times were closely allied in type to the tarpan. Perhaps most characteristic of rural Britain today is the heavy, patient carthorse, frequently weighing a ton or more, with sleek coat of varied colour, broad back and a distinguishing growth of long hair or "feather" round the lower parts of the legs and hanging over the hooves. The height of horses is measured in 'hands' of four inches, at the shoulder, and a good carthorse may stand 17 or 18 hands—5 feet 8 inches or 6 feet, so that it requires a very tall man to see over the animal. It is perhaps more dignified to refer to the Shire Horse—a reminder that breeding is characteristic of the grassy Midland shires of England—than to a 'carthorse', but even this designation is a poor tribute compared with such a title as the Great English

War horse. For such was the origin of the breed: when both horse and rider carried heavy armour a really strong animal had to be bred for the purpose and speed was sacrificed. (See Plates XVII and XVIIIb.)

In Scotland the Shire horse gives place to the somewhat lighter built Clydesdale and only two other agricultural work horses need to be mentioned as surviving in Britain. One is the Suffolk Punch, the other a recent comer from France, the dappled grey Percheron. The latter proved so valuable to the Artillery for heavy work in the First World War that numbers were introduced as farm animals.

It may perhaps be claimed that James I, if not the first to import Arab horses into Britain, at least founded one of the first Arab studs. Although this stud was later sold and dispersed, Cromwell as a countryman also had an eye for a good horse. He realized that the days of heavy armour and the heavy horse able to bear the weight of its own armour and an armoured rider were over. He imported a grey Arab stallion, long famous through the country under the name of the White Turk. Charles II loved horses as he loved other good things in life, bought Arab horses freely and established another stud from which stem many of our present-day thoroughbreds. Thus it may be said that the reign of Charles II saw the rapid spread in popularity of the light swift animals of Arab stock. Thus the English Thoroughbred or racehorse is of almost pure Oriental blood but through breeding has been separated from the Arab and is superior in size, weight, speed and stamina.

No account of the horse in rural Britain, however brief, would be complete without reference to the semi-wild breeds of ponies so long associated with Dartmoor, the New Forest, the Western Isles, Shetlands, Scottish Highlands and elsewhere in the hill-lands. Shetland ponies, smallest of all, and generally under 40 inches or 10 hands high, were much appreciated as children's mounts in the days of large country houses, large families, nursemaids and grooms and low taxation. Dartmoor ponies breed on the moors, come down to the surrounding villages for food in the hard winters and were formerly destined to spend the rest of their days in coal mines unless pensioned off for long and meritorious service and a final trip to the surface. Some ponies remain but as a source of income to Dartmoor farmers they are no longer significant. (See Plate XVIIIa)

The various pony breeds of Highland Britain are believed to be descended from the Celtic Pony* of the Western Isles and still found in

*Older authorities named this *Equus celticus,* but the modern view is to regard it as a race of the domestic horse *E. caballus.*

fairly pure form on the Island of Barra. Closely related are the heavier
Highland Pony, the Fell Pony of the Lake District, the Dales Pony of the
Pennine Dales and the Welsh Pony of the hills of Wales. All these—
especially in Scotland—are used as pack animals notably for the carrying of peat.

TURKEYS

Every year on the eve of the last Thursday in November millions of
Americans may be found making the annual trek home for the great
family festival of Thanksgiving. The Pilgrim Fathers who landed on the
shores of New England in December, 1620, had experienced their first
New England winter, so much more severe than anything they had ever
known at home in England that many of their number perished. In
the spring they had laboriously cleared and tilled patches of land and
watched anxiously for their first crops to come to harvest. Though they
gave thanks to God for that first harvest, largely of maize which
friendly Indians had taught them to grow, it is doubtful whether many
would have survived the first winter had it not been for the wild turkeys
of the woods. A less practically-minded nation might have made the
turkey a sacred bird, thenceforth free to wander its native woods safe for
ever from the huntsman's rifle. Instead the turkey hatched, reared, and
fattened for the occasion ; variously stuffed, roasted and embellished
with such *embarras de richesse* as sweet potatoes and marshmallow sauce is
destined to be eaten in perpetuity on the anniversary of the first Thanksgiving. The turkey is America's one main contribution to the domestic
fauna of the Old World. In the wild state it thrives in the bitter snowy
winters and hot summers of its native woods : as a domestic bird it is
peculiarly delicate and prone to fatal chills and ills. As a result turkey
farming is a specialized and precarious form of farming, though highly
remunerative when a flock is safely reared. The turkey farms of Norfolk
and other parts of East Anglia and those of Northern Ireland provide a
large proportion of home-reared birds : elsewhere the farmer's wife may
rear a few in the hope of a bigger return than from ordinary fowls if they
survive the requisite period. None can deny that the turkey is a noble
ornament to the farmyard, large enough and weird enough (Plate 23b)
to strike the imagination of every child and with a fascination enhanced
by the curious gobbling noise. Some even claim that this noise,
resembling *turk, turk, turk*, is the origin of the name.

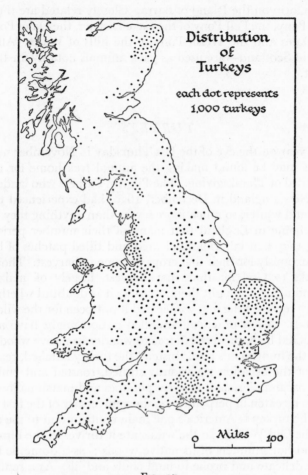

Distribution
of
Turkeys

each dot represents
1,000 turkeys

0   Miles   100

FIG. 27
The Distribution of Turkeys, 1950
Small numbers are widely distributed but specialized turkey farming
is almost restricted to East Anglia

Plate XIXa. (*above*) LOMBARDY POPLARS LINING AN ARTERIAL ROAD, CHESHUNT
Giving a very Continental appearance to the English scene (see page 176).
(*John Markham*).
b. (*below*) AN AVENUE OF HOLM OAK, HOLKHAM, NORFOLK
A Continental introduction, *Quercus ilex* (see page 176). (*John Markham*)

a.—THE MONTEREY CYPRESS AT HOME. *Cupressus macrocarpa* survives with difficulty in its native home—an old group at Cypress Point, Monterey, California (*p. 181*)

b.—THE MONTEREY CYPRESS IN ENGLAND

Cupressus macrocarpa flourishes in the milder parts of Britain and has become a familiar part of the village scene. This well-established specimen was in the garden of the author's former home, Bude, Cornwall. It was cut down on the insistence of the nationalised Electricity Supply Board that it interfered with the antiquated overhead electric wires (*p. 181*)

(Photographs by L. Dudley Stamp)

Plate XX

The story of the Pilgrim Fathers and the New England turkeys is familiar to every American schoolboy but the true facts of the origin and introduction into Europe of the turkey may never now be known with accuracy and precision.

In the first place the American turkey was confused by early writers with the African guinea-fowl, and it is sometimes uncertain to which bird sixteenth century references apply. The guinea-fowl became known to Europe through the Mahommedans of North Africa, then familiarly called "Turks", so that the names "Turkey-cock" and "Turkey-hen" were applied to birds of West African origin. The turkey as we know it is *Meleagris gallopavo*, one of two species of an exclusively American genus. It ranges wild from Colorado to Pennsylvania, formerly from southern Ontario to Mexico and of five local races one, the Mexican subspecies, would seem to have been domesticated by the Mexican Indians before the days of Columbus. It is sometimes said to have been introduced into Spain as early as 1498 whence it spread rapidly as a domestic bird through Western Europe, being imported into England between 1524 and 1541. Others claim that the importation of the turkey was one of the immediate practical results of the Cabot discovery voyages of 1497-98. Certainly by 1555 there are references to its use on English tables, whilst Tusser writing in 1573 commends it as Christmas fare. Certainly also by the time of the Pilgrim Fathers it had become a much-prized domestic bird in Europe. Dr. R. H. Brown* asserts that it "actually made the American-bound voyage with some of the first colonists. The turkey was but one of the many useful native products that were introduced in this roundabout fashion at the time of colonization." Since contemporary or early accounts refer to the abundance of game, especially wild turkeys and deer, it may be that these introduced European birds, if the account be true, were crossed with the wild ones. Certain it is however that in the autumn of 1621 the Governor (says Winslow, in Mount's *Relation*), "Sent four men on fowlings, that so we might after a more special manner rejoice together after we had gathered the fruit of our labours."

In England there seems to have been little change in the turkey for some three centuries. In 1865 the original Mexican stock was known as Cambridge Bronze. About that time breeders developed the Norfolk Blacks and a buff variety or "Fawns". Birds of both sexes of the old Bronze reached 15, 16, 17 and 18 lb. respectively at 24, 26, 28 and 30

Historical Geography of the United States. Harcourt Brace, New York, 1948, pp. 35-36.

weeks, and old tom birds weighed up to 40 lb. Smaller families led to a demand for smaller birds and after about 1930 broad-breasted birds were developed averaging 12 to 14 lb. dressed weight. Thus the recent evolution of the turkey is in line with the demand for smaller joints of beef and mutton.

The map showing the distribution of turkeys in Britain brings out a number of interesting points.

THE GUINEA-FOWL (*Numida meleagris*)

The Guinea-fowl is remarkable as one of the very few contributions of the African continent to the modern farmyard. It is a native of the whole of Africa south of the Sahara, but it was probably from the Guinea coast from the Gambia to Gabun that it was reintroduced into Europe by Portuguese African explorers about the early part or middle of the 16th century.* It has in fact undergone little change with domestication since first described and figured by Gesner (*Paralipomena* in 1555) and Belon in the same year. The word 'reintroduced' correctly states the position because the bird was known to and domesticated by the Romans, being known to them as the *Gallina africana* or *numidica,* and was the *meleagris* of the Greeks (μελεαγρίς).

Considerable misunderstanding has arisen because Gesner, Belon and even Ray confused this bird with the turkey and Linnaeus, by using the generic name *Meleagris* for the turkey, seems to have fallen into the same error.

There is little difference between male and female birds and both are ornamental denizens of the farmyard. In the wild state guinea-fowl, though heavy birds, have considerable powers of flight and the domestic birds, left to themselves, will roost normally in trees.

POULTRY

The word poultry is commonly applied to domesticated edible birds —fowls, turkeys, guinea fowl, domestic ducks and geese. In contradistinction the word game is used for edible wild birds (and animals).

The domestic fowl in all its numerous varieties is now regarded as derived from the jungle fowl of India, *Gallus gallus,* which is still com-

*Darwin (1868, vol. I p. 294) however traces the descent of the domestic form not from one of the west African races but from *N.m. major* (which he called *N. ptilorhynca*) of arid *east* Africa.

mon in the woodlands of India and neighbouring countries. Indeed the domestic fowl of Indian villages differs so little from the wild bird as to be almost indistinguishable. On one occasion when I was travelling by bullock cart in the jungle of upper Burma, I had an exciting time catching a pair of jungle fowl with the help of my boys and well-aimed, if primitive, weapons, only to discover that they were my own domestic fowl, escaped from the familiar fowl basket on the bullock cart. The jungle fowl, of which the cock and hen differ greatly, are not unlike the game birds formerly kept for cock-fighting in Britain and perpetuated today for show rather than for meat or eggs.

Darwin distinguished thirteen principal breeds : modern poultry fanciers list over a hundred.

DUCKS

The various breeds of domestic duck in both Europe and America are all derived from the Mallard or common wild duck (*Anas platyrhynchos*) which has a very wide range in the northern hemisphere, reaching from as far south as Panama, Egypt, India and Borneo in winter to well within the Arctic Circle in summer. It is not known when the duck was first domesticated and even today the line is not sharply drawn between wild and domestic ducks. On ponds and quiet streams near towns wild ducks appear and expect to receive most of their food from human hands : in the open farmyard a pair of ducks may disappear in spring to return later with a brood of ducklings.

In its truly wild state the mallard is strictly monogamous and a pair may mate for life : under domestication the drake becomes notoriously polygamous. In the wild the couples pair very early but the ceremonies of courtship take a considerable time. The nest in dry grass under a bush, in a hollow tree or at the top of a pollarded willow is lined with down which the female pulls from under her breast feathers, and which can later be arranged as a sort of coverlet for the nine to eleven pale greenish eggs on the rare occasions when she leaves the nest. The male assists in guarding the nest till the young are hatched. Thence forward all the work falls to the female, including the all-important first introduction to the water (sometimes involving a considerable and very secret journey) and she stays with them till they are fully fledged. Whilst the ducklings are still unable to fly both birds undergo a moult which robs them of the power of flight for about three weeks.

Duck shooting is still a popular sport and wild duck, small but tasty, form the basis of more than one menu for gourmets. Before the draining of the great fenland areas large numbers of duck—together with geese and various game birds—were caught by professional and amateur fowlers. A regular system of secluded ponds with tame decoy ducks was used and the evidence still remains on Ordnance maps of "decoy ponds" and "decoy woods". A few are still worked in Britain. In this way the past has left once again its mark on the present day countryside.

Duck eggs have a stronger flavour than ordinary hen eggs. Largely for this reason domesticated ducks are kept primarily for their meat. They mature quickly and some breeds reach 5 to 7 lb. in weight in 10 to 12 weeks. Probably the best known of the British breeds is still the Aylesbury Duck, though the once famous industry in the Vale of Aylesbury itself has fallen into insignificance. When the Rev. John Priest wrote the *General View of the Agriculture of Buckinghamshire* (1810) he described how this white duck was bred by poor labourers and cottagers in their own homes—often crowded for shelter at night in the single bedroom or living room of the cottage—and not by farmers. Ducks like to be fed in water and the labourers were accustomed to dig a small pond with thatched shelters for daytime use. It is now known that ducks kept away from water fatten more rapidly.

GEESE

The domestic goose is derived from the bird commonly called in English the Grey or Grey-lag Goose (*Anser anser*). The wild bird has a very wide range in the Old World from Scotland and the extreme north of Europe in Lapland to Spain, the Balkans and eastwards into China. As Darwin commented long ago, its domestication is of very ancient date and yet scarcely any other creature which has been domesticated so long and has been so extensively bred in captivity has varied so little. Domestic geese have increased greatly in size and fecundity and have usually lost the browner and darker tints of the wild bird, often being wholly white. Even that change is very ancient, for the Romans held white geese in special regard : indeed the Sacred Geese of Rome which warned the sleeping guards of an impending assault were white. The very words goose, from the Anglo-Saxon *gos*, and gander from the Anglo-Saxon *gandra* are indicative of early interest in Britain. *Anser anser* is the only species of wild goose indigenous in the British Isles and until the draining of the Fens bred freely in England as it still does in remoter parts of Scotland. One of the reasons for opposition to the draining of the Fens was the threat to the breeding of wild geese, as the young were caught in large numbers and added to the flocks of tame geese. The name "greylag" was suggested by Professor E. W. Skeat to have been derived from the fact that this goose lagged behind (cf. laggard or loiterer) when all others left for their northern breeding quarters.

In contrast to ducks the sexes are alike and though the nesting habits of geese resemble those of ducks, the males remain and take more part in family duties after the hatching of the young. Like ducks geese have a summer moult at the breeding season ; and lose their power of flight. Geese when flying generally do so in a V-formation.

The breeding of tame geese has played an important part in British country life, especially in some areas, at least since Anglo-Saxon times. In such eastern counties as those bordering the Fens, flocks of a thousand or more were tended by a gooseherd or gozzerd, just as sheep are tended by a shepherd. They were taken regularly to pasture and to water, they were plucked for their yield of down and feathers four or five times a year—incidentally new plumage was usually white whereas previous may have been partly brown—and in autumn were driven to market. Travelling at about a mile an hour they could be made to cover about ten miles a day, though one wonders whether the flesh of birds which had thus made the journey from the Fens to London was as tender as one might require it today. Selected birds were kept and fattened, and a fine fat goose was the traditional Christmas bird—as Dickens reminds us —rather than the exotic turkey. The flesh of the turkey resembles closely that of the ordinary fowl, whereas that of the goose is particularly distinctive. Two modern breeds, the grey Toulouse and the white Embden, have become outstanding meat-producing breeds. Commercial goose-farming is more important today in Germany, Austria and parts of France than in Britain, and in France special feeding is sometimes practised in order to secure enlargement of the livers for the production of the favourite *pâté de foie gras*. In England small numbers of geese are often kept as a sideline : they are omnivorous feeders and eat off grass in wet land but they tend to make such land foul after a short time with their droppings.

CHAPTER 12

POMONA: THE ORCHARDS OF BRITAIN

Rural Britain owes not a little of its scenic charm to its fruit orchards. Especially is this true in spring when well-kept orchards are seen in full blossom. Few sights surpass in tranquil beauty that of a Kentish cherry orchard of well-grown mature trees laden with pink-tinged white blossom and with well-fed, well-fleeced Kentish sheep grazing serenely on the lush growth of spring grass. Such a picture is satisfying aesthetically but it satisfies also because it is the result of a well kept and a wisely used land. The very mention, too, of a Kentish cherry orchard is a reminder that fruits are strongly localized in their distribution and so help greatly to establish the strong individualism of the different parts of the country. It is true that few parts of the country-side lack their cottage gardens with a few apple or plum trees but fruit orchards on a scale to characterize or dominate the landscape are definitely restricted to certain areas.

The map showing the distribution of orchards in Britain based on official statistics is supplemented by the table of orchard acreages in a recent year, taken also from official statistics. Certain areas stand out clearly and pride of place on any count belongs to the orchard county of Kent, with approximately a quarter of its non-grass farmland under fruit. Second is the Evesham district of Worcestershire. The famous old cider counties of Hereford, Somerset and Devon show widely scattered orchards. The major patches in eastern England are largely later developments in commercial fruit production—often, as in Cambridge-shire, associated with small fruit. Localization is still more marked if one takes the different fruits. Most widespread as might be anticipated, are apples. Even when considerable quantities of perry were made from pears, the cultivation of pears was never as widespread. Plums of different types are grown in most parts of Britain, but on a commercial scale are especially characteristic of Worcestershire. Cherries definitely

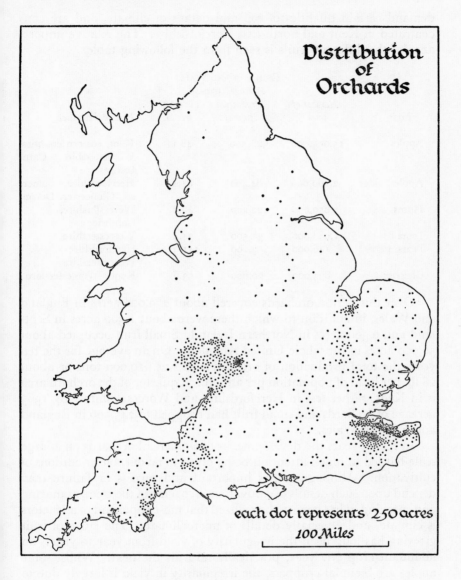

FIG. 28
The Orchards of England and Wales

demand certain qualities of soil and details of climate, and are concentrated in Kent and north-east Herefordshire. The relative importance of the different fruits is seen from the following table.

ENGLAND AND WALES
Average 1939-48

Fruit	Number of trees	Production in tons	Yield per tree lb.	Principal counties
Apples	15,055,000	322,300	48.1	Kent, eastern counties, Worcestershire, Cambridgeshire
Apples, cider	2,360,000	84,400	81.4	Herefordshire, Somerset, Gloucester, Devon,
Plums	5,859,000	122,200	46.8	Worcestershire, Cambridge
Pears	1,988,000	30,500	34.4	Worcestershire
Pears, perry	146,000	7,200	111.9	Herefordshire Worcestershire
Cherries	835,000	20,000	53.6	Kent, Worcestershire

Over this period orchards covered about 270,000 acres in England and Wales, in addition to which there were about 1,300 acres in Scotland and 9,500 acres in Northern Ireland. Small fruit occupied about 50,000 acres in the whole United Kingdom. On an average for the ten years the total production of orchard fruit was 619,000 tons or about 28 lb. per head of population per annum. A quarter of the orchard area is in Kent, rather less in Herefordshire and Worcestershire. By 1963 acreage of orchards and small fruit had dropped to 245,500 in England and 2,600 in Wales.

In recent years the development of commercial orchards on a large scale has led to much attention being given to favourable conditions of cultivation. Although soil is important and success or failure may depend upon such details of cultivation as spacing of the trees, manuring and spraying, it has come to be realized that most important of all factors is climate, and especially details of microclimate. The curse of fruit growing has long been the irregularity of yield from year to year: an abundant crop one year, practically no fruit the next. Whilst some apples are biennial croppers, the irregularity in yield is largely due to spring frosts. If frosts occur after blossoming and before the fruit has 'set' there must follow a crop failure. If blossoming is delayed by a cold spring but is not followed by frost, the chances of a good crop are excellent. There are thus advantages in choosing sites (such as some

F. Fraser Darling

Plate 9. SCOTTISH BLACKFACE SHEEP

In their home area—the Scottish highlands, Loch Sunart, Argyll   *(p. 141)*

Plate 10a (above). PEDIGREE WILTSHIRE RAMS. Showing the characteristic absence of wool on this very old breed   (*p. 144*)

b (below). WELSH MOUNTAIN SHEEP. In a lambing enclosure on a mountain farm (*p. 143*)

(*Photographs by L. Dudley Stamp*)

ACREAGE IN 1949 AND 1963

England	1949 Orchards with grass or crops 253,686	1949 Orchards with small fruit 13,081	1963 Orchards and small fruit 249,400
Bedford	1,720	19	1,100
Berkshire	2,996	85	2,300
Buckinghamshire	3,174	98	2,100
Cambridge	4,873	372	4,000
Isle of Ely	3,455	1,114	5,900
Chester	1,299	133	900
Cornwall	2,771	131	1,700
Devon	19,841	421	12,500
Dorset	1,916	30	900
Essex	10,930	603	14,600
Gloucester	14,711	227	11,200
Hampshire	2,474	162	2,900
Hereford	23,748	266	21,100
Hertford	1,721	63	1,300
Huntingdon	1,708	312	2,400
Kent	71,258	3,465	78,900
Lancaster	1,320	154	600
Monmouth	2,473	85	1,300
Norfolk	8,071	1,929	14,700
Nottingham	1,866	81	1,300
Salop	2,759	39	1,900
Somerset	18,638	164	13,000
Suffolk, East	2,822	214	} 9,100
Suffolk, West	2,773	110	
Surrey	1,500	96	1,300
Sussex, East	3,706	284	5,100
Sussex, West	2,262	116	2,400
Warwick	2,984	132	2,400
Worcester	24,926	1,006	21,500
Wales	4,128	186	2,600

All other counties have less than 1,000 acres each in 1949 but Lincoln had risen to 3,400 in 1963. Notice the use of the sunny eastern counties and the fall in the old cider counties of the west.

north-eastern slopes) where blossoming will be delayed. This is contrary to previous belief in the advantage of sunny south slopes. But even more important is the fact, so ably demonstrated by a practical fruit grower, Raymond Bush, that cold air behaves much like cold water. It will pour down a hill slope at night from a cold plateau above which acts as a gathering ground. Thus a slope with no reservoir of cold air above is important. If the cold air meets an obstruction it will pile up as water in a lake and form a frost pocket. Hence free drainage *below* an orchard is important and a wind break or wood on the downward side from an orchard may be fatal instead of beneficial. Free drainage to tide water is particularly valuable because of the ameliorating or steadying influence of bodies of sea water. The principles are really quite simple but they have not been understood in the past. If an annual crop, such as a cereal, fails it need not be repeated by the farmer but he hesitates to pull out an orchard which has taken ten years and much capital to establish. It has been stated with considerable jusification that at least half the orchards of Britain are wrongly sited and ought to be replaced. Even some well known experimental stations are sited on land which forms a frost pocket—with fatal results—in most years in the spring. The behaviour of cold air as it affects orchards is illustrated in the accompanying diagrams. The effect of cold air spilling over a wall is one which should interest many home gardeners (see Plate 12).

The commonest orchard fruits—apples, pears, plums and cherries—all belong to the *Rosaceae*. Of each there are representative wild trees native to Britain, though the cultivated varieties are not necessarily derived directly from the surviving wild stocks. Almonds, peaches, apricots, nectarines, medlars and quinces are also members of the *Rosaceae* but derived from species or genera not native to Britain.

In its innumerable varieties, running into hundreds, the apple is the most widely cultivated of all fruit trees. It seems to be certain that all varieties are derived from the wild crab-apple, known now as *Malus pumila*, formerly as *Pyrus malus*, though cultivated apples date from pre-Roman times and have been a feature of the British countryside certainly since Norman times. Though never in pure strains the wild apple is found widely distributed in woods through Britain except northern Scotland : it does not develop beyond a small tree from 20 to 50 feet high of irregular growth. The wild crab apples, yellow or red in colour, are borne on long stalks and because of a high content of malic acid remain sour even when ripe. Like so many of our wild fruits they were formerly much more appreciated than they are now—for the making of crab apple jelly and for the embellishment when roasted of hot punch. There are numerous cultivated varieties of the crab apple grown in gardens and parks for the sake of their abundant spring flowers and

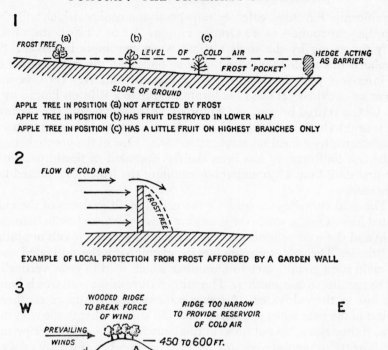

APPLE TREE IN POSITION (a) NOT AFFECTED BY FROST
APPLE TREE IN POSITION (b) HAS FRUIT DESTROYED IN LOWER HALF
APPLE TREE IN POSITION (c) HAS A LITTLE FRUIT ON HIGHEST BRANCHES ONLY

FIG. 29
Diagrams to show the behaviour of cold air causing frost damage

sometimes for their attractively tinted foliage in autumn, but they
have scarcely become dominant factors in the country landscape. It
is otherwise with the innumerable named varieties of the apple cultivated for fruit. The days when a farmer supplemented his ordinary
income by selling apples from a dozen different varieties, often unnamed,
in his orchard have passed and given place to the specialist production
from large scale orchards. Custom differs as to the spacing of trees and
whether the ground between should be grassed (to be cut or grazed)
or cultivated but the marked tendency is for only two or three varieties
to be grown. No apple surpasses in popularity or general utility that

magnificently flavoured eater, equally good as a cooker, which will also keep the year round—Cox's Orange Pippin. A ripe Cox's betrays itself ready for eating by the seeds rattling when the apple is shaken. So popular is this apple that more than one-half of all the trees planted in the inter-war years in Britain were of this one variety. Yet Cox was almost an accident. It was raised from the pip of a Ribston Pippin by a Mr. Cox, a retired brewer, at Colnbrook near Slough, and the original tree is said to have survived till blown down in 1911. It was introduced to commerce by a local nurseryman in 1850. One of the great changes of the last half century has been the development of 'bush' orchards (the first date from 1890 and 1900) enabling the fruit to be picked far more easily.

The wild pear *Pyrus communis* is the origin of all varieties of the cultivated fruit. On the whole the pear demands a warmer climate than the apple and the wild pear occurs only occasionally in the woods of southern Britain. The wild tree resembles in general habit the cultivated one, the main stem giving place to a number which tend to grow vertically and so parallel to one another. The spiny nature of the wild tree is however lost in the cultivated. Another indication of the more southern habitat of the pear when compared with the apple is that in the wild the fruits, if they ripen, do not do so until about November. When ripe the flesh is gritty to the taste because of the large number of woody cells or idioblasts which are absent from the flesh of the apple, and which it is the aim of the grower of cultivated pears to eliminate entirely. Although cultivated pears are found in gardens in many parts of Britain, on a large scale pear orchards are limited to the South and the southern Midlands : Hertfordshire, Kent, Worcestershire and Norfolk. Before the great spread of suburban London rural Middlesex was noted for its pears.

In shape the hairy-fruited quince (*Cydonia oblonga*) somewhat resembles a pear. It has no native relative in Britain and in cultivation is now rare ; it ripens only in southern England. Quince jelly was once a typical country preserve and a single quince added piquancy and a peculiar fragrance to apple pie. The medlar (*Mespilus germanica*), also allied to the pear and introduced from continental Europe into Britain about the end of the 16th century, has now become rare.

Several wild species have contributed to the many varieties of cultivated plum. The blackthorn or sloe (*Prunus spinosa*) in many areas is almost as abundant as the whitethorn, hawthorn or may (*Crataegus* spp.) as a native hedge shrub and enormous numbers were planted as 'quick set hedges' (i.e. hedges of shrubs set live or quick) in the days of the enclosures. Country children formerly rather than now consumed quantities of the wild sloes which become sweet when dead ripe, though very sour and astringent earlier. Though it grows to a small tree without any pronounced main stem, most blackthorn is seen in hedges where a close mat of spiny branches renders it particularly valuable. It may start to blossom as early as March and is often the earliest reminder of an incipient spring. But hedgerows are part of the artificial or unnatural landscape of Britain and the sloe should be classed strictly as one of the frequent rather than dominant small trees of woodland margins—the constant companion there just as it has become in hedgerows of the hawthorn.

Another British wild plum which may or may not be a native is the bullace or bullen (*Prunus insititia*)—found as a more robust and rather uncommon tree in open woodland. The bullace is cultivated in country gardens and gives an abundant yield of small yellow plums : it is possible that the wild form is an escape. In its fruit the damson resembles a very large sloe : something of the sharp flavour of the sloe remains. It has been claimed that the damsons and other small sharp flavoured plums cultivated in the Kendal area and elsewhere in the north are popular in northern industrial towns because an atmosphere laden with smoke and acid fumes has dulled the palates of the inhabitants to an appreciation of more delicate flavours.

Most plums are however derived from the wild plum (*Prunus domestica*) found rather rarely in woodlands and distinguished by an absence of spines. This species is believed to have originated in Asia Minor and may be regarded as only naturalized in Britain.

No less than three wild cherries are found in Britain and they may all be native. The commonest is the gean (*Prunus avium*), the others are the bird cherry (*P. padus*) and the dwarf cherry (*P. cerasus*). Probably cultivated cherries are derived some from *P. avium* and others from *P. cerasus*, and it is believed that as cultivated trees all cherries were introduced into Britain from continental Europe. There are records of the sale of cherries in London in the early part of the 15th century, but it was the importation of tree stocks from Flanders in 1533 under orders from Henry VIII which established both the popularity of the fruit and the preeminence of Kent as the cherry orchard county.

Hops

Though Britain has long had beer, it is only for the last four centuries that British beer has derived at least some of its tang from an infusion of hops. The crop was introduced from the Netherlands probably about the middle of the 16th century (though some authorities say the end of the 15th), and is noted as spreading rapidly by Tusser in 1551, but not mentioned by Fitzherbert in 1523. The introduction was into Kent and, though hop gardens spread over much of southern and midland England, they remain characteristic of Kent where over half the present day total acreage is found. The other main centre is in Herefordshire and neighbouring parts of Worcestershire. The acreage has dropped greatly from the eighteen-eighties: the crop makes severe demands on soil, capital and labour but has certainly played an interesting role in moulding scenery and social life in parts of England. Hop soils must be strong, deep, well drained loams and hops compete with fruit. The old system was to train the vines up poles of chestnut or ash 12 or 16 feet high, and connected by strong coir yarn. The provision of hop poles was a most important objective in the management of many coppiced woodlands in the south-east. The modern system is a permanent framework of stout posts and galvanized wire, with coir yarn strings from the wires to the ground renewed annually. Hop plants are male and female: about one male to every hundred females is planted. By about June 21st the vines reach the top of the strings and the females fall over to form a clustered head. By September all is ready for the arrival of the hop-pickers—casual labour from the towns. Hop picking in Kent provided a fortnight in the country for families of Londoners—father, mother and children—for generations. Hop-picking by hand virtually came to an end in 1963. The kilns or oasthouses used for drying the hops are a characteristic feature of Kentish scenery and testify to the former much wider spread of hop cultivation. The older oasthouses were round with picturesque white cowls which turn from the wind; the more modern are built over square floors; practically all are now obsolete.

CHAPTER 13

SYLVA: A DISCOURSE ON TREES*

FEW PEOPLE realize the poverty of the British Isles in truly native forest trees. Provided a species was introduced sufficiently long ago it is accepted by those who dislike the spread of exotics, especially exotic conifers. Few would wish to banish the chestnut or the sycamore because they are not native, few would deny the beauty of young larch foliage in spring despite the fact that it is an alien. We now seek strenuously to preserve the clumps of beeches planted on the heights of the downs, as at Chanctonbury Ring, despite the fact that they are artifical creations deliberately designed to alter the natural landscape. Yet we often hear decried the 'ornamental' trees of the great parks and the introductions of landscape gardeners. Still more curious is the condemnation of those who seek to restore the beauty of the noble forest to areas laid waste by the woodsman's axe in the Middle Ages and since occupied by almost worthless bog and moor. Dr. Johnson saw in the Western Highlands only depressing desolation, older agricultural writers invariably stigmatize moorland and rough grazing as "waste" and to William Cobbett London's beloved Hindhead Common in Surrey was country "the most villainous God ever made". Today many regard it their life's work to prevent the restoration of Britain's natural forest mantle to those poor naked hills in which they find their current concept of beauty. True, their opposition is lessened if "native" trees are to be planted, ignoring the fact that the geological accident of Britain's early separation from the Continent after the Great Ice Age is alone responsible for the very short list of indigenous British trees. It was this accident alone which prevented such trees as the Norway spruce and European larch from reaching Britain without the aid of man. Indeed the Norway spruce had once flourished in Britain but was driven out by the later onslaught of the re-advancing ice.

*My thanks are due to Professor Sir H. G. Champion for valuable comments on this chapter.

It is only of recent years that research has established without doubt the right of certain trees to be called "native". For long it was contended that the beech had been introduced by the Romans and its restricted range—widespread only in southern England—was regarded as evidence of its introduction from the south-east. It is difficult to identify with certainty fragments of wood from peat beds : leaves and fruits have rarely been preserved sufficiently well for the species to be named with certainty. A great advance was made when peatbeds were searched for seeds and the hard seed cases identified. This method of study is associated especially with the names of Mrs. E. M. Reid (whose husband Dr. Clement Reid was for many years an officer of the British Geological Survey and himself very interested in the development of the British flora) and her able assistant Miss M. E. J. Chandler. Mrs. Reid had to form a reference collection of seeds before she was able to identify many of those found fossil in interglacial or late glacial deposits. Naturally the seeds of many plants were not preserved and those found were of species of plants which at the time were growing near the peat bogs or waters in which deposits were being formed. Nevertheless Mrs. Reid and Miss Chandler were able to demonstrate fluctuations in climatic conditions in late glacial Britain.

A very great advance was made possible by pollen-analysis briefly described in Chapter 1. Pollen grains are disseminated by wind—the bulk short distances only but the finer grains from fir, larch, spruce and beech may carry up to 15 miles, elm to 60 and alder to 80. Though the forest may have been some distance from the bogs or ponds in which the pollen-rain fell, we can still build up a good general picture of the vegetation over a wide area.

The latest of the pollen-analysis stages (VIII) of Godwin corresponding with the Sub-Atlantic Period of Romano-British times gives us the picture of Britain's plant cover at the time of the first Roman reconnaissance invasion of B.C. 55-54. (see p. 10).

The natural vegetation of the greater part of Britain at that time, untouched by man, was undoubtedly forest. We cannot be certain of the upper limit but there is evidence (as on Dartmoor) that in the south of the country it extended up to 1,250 or even 1,500 feet. In the north and in Scotland the limit was naturally lower just as it is at the present day. A great sea of hardwood forest, in which oak predominated stretched over most of the lowlands giving place to coniferous or mixed forest with Scots Pine (*Pinus sylvestris*) on higher ground in the north or on.

Plate 11a (*above*). DEVON LONGWOOL SHEEP. Gathered for shearing in the field, showing the ringlets of wool (*p. 147*)

b (*below*). DOWN LAMBS. Folded on a rich ryegrass pasture on the chalk downlands of Wiltshire (*p. 144*)

(*Photographs by L. Dudley Stamp*)

Plate 12. MICROCLIMATOLOGY OF A BRICK WALL

This wall faces south and cold air pouring down from higher ground to the north
has completely withered the branches of the fig tree which project above the wall,
whereas the fig leaves remain unharmed where protected by the wall   (*see fig. 29*)

sandy soils elsewhere.* On calcareous soils oak was associated with the ash or gave place to beech woods : birch and hazel were widespread on the flood plains of rivers and in swampy carrlands oak was replaced by the more easily cleared alder : a circumstance which helped to attract early settlers to riverside settlements.

A number of the trees now known to be native to Britain do not normally occur in close stands forming forest. They seem rather to flourish in situations where they can enjoy fully surrounding light and air. An outstanding example is the common elm (*Ulmus procera*), now the commonest hedgerow tree but rarely found in woods. The wych elm (*Ulmus glabra*) is undoubtedly a native but the status of the ordinary elm which probably does not grow wild on the Continent, is still in doubt. Elm pollen does not permit identification as to species. The native lime (*Tilia cordata*) is also a tree associated with avenues and open situations rather than woodlands. Only occasionally does that tree so characteristic of British churchyards, the yew (*Taxus baccata*) occur as woodlands though a few yew woods, notably at Boxhill in Surrey, are famous. The rowan or mountain ash (*Sorbus aucuparia*) tends to grow in isolation and reaches a higher elevation than any other British hardwood tree. The holly (*Ilex aquifolium*) though found in woods has come to be regarded rather as a common hedgerow tree, and is another native. Widely scattered in woodland though rarely numerous are aspens (*Populus tremula*), wild cherry (*Prunus* spp.), wild pear (*Pyrus communis*) and wild apple (*Malus pumila*).

Turning now to aliens, the Romans are commonly credited with the introduction of the Spanish chestnut (*Castanea sativa*), the sycamore or great maple (*Acer pseudoplanatus*), walnut (*Juglans regia*) and mulberry (*Morus nigra*). As soon however as one attempts to seek accurate data concerning times and circumstances of introductions the difficulties are innumerable and often insuperable. Some authorities, for example, say that the sycamore and mulberry were not introduced by the Romans and did not reach Britain till the 15th century.

Even in quite modern times there may be no written or published record of a first introduction ; long after the event someone's statement made from memory or hearsay information may be recorded and subsequently accepted by one writer after another. With some of the older introductions there may be confusion as to the identity of the tree. In

*It is important to note however that the Scots Pine which has colonized much of southern Britain of recent years is an introduced continental form (personal communication from Sir Edward Salisbury).

the case of fruit trees such as the apple, pear, cherry and plum, origins are particularly difficult to trace. Are the "wild" species the original stocks or are they escapes ? Thus the supposedly Roman introductions must be accepted with reserve. All that one can say of the Spanish or Sweet chestnut (*Castanea sativa* or *vesca*) is that it is a native of the Mediterranean lands of southern Europe and has long been grown successfully in Britain ripening its fruit as far north as southern Scotland. The early introduction of such a valued food-bearing tree is understandable: it is more difficult to accept very early dates for the introduction of trees which are purely or mainly ornamental. The sycamore and sweet chestnut may both prove to be natives.

The Horse chestnut (*Aesculus hippocastanum*) affords an example of a greatly appreciated ornamental tree but of which the nuts despite their attractive shiny red brown skins have never been widely used and are commonly regarded as injurious to man and animals, even poisonous. It is certain that both horses and pigs refuse to eat the nuts and exhaustive experiments during the Second World War failed to produce a meal of any nutritive value from them. The tree has the advantages of growing rapidly to an attractive pyramidal shape : the tender green leaves which burst from the sticky buds are the very epitome of spring : in May or June the great inflorescences in white and pink will bring visitors many miles to see a chestnut avenue whilst in October few small boys can resist collecting a pocketful of "conkers"*. The tree itself is believed to be a native of Greece : when it first arrived in Britain is uncertain but probably was some time in the sixteenth century. See Plate 24b.

Britain in general, her great towns in particular, and above all the heart of London would be greatly the poorer without those magnificent aliens the plane trees. The oriental plane (*Platanus orientalis*) is a native of Greece and western Asia and was introduced by the Romans into south-western Europe as a shade tree, valued for its large translucent five-lobed leaves and its picturesquely peeling bark as well as its good-tempered tolerance of cutting, pruning and training into many forms. The plane was probably brought to England in the sixteenth century. A closely allied species is *Platanus occidentalis* introduced into England in

---

*The popular country game of "conkers" consisted of boring a hole in a nut, holding it suspended by a piece of string some 18 inches long threaded through the hole and knotted, whilst it was hit by the opponent similarly armed. A well directed blow would split the one nut, leaving the other the conqueror—hence conker. As the season passed one tough specimen might survive many such battles and would be treasured long after the conker season till thrown out and burnt by an assiduous spring-cleaning mother.

the seventeenth century from America where it flourished in the basin of the Ohio and parts of the Mississippi under the name of sycamore or button wood. There are some reasons for believing that the London plane (sometimes distinguished as *P. acerifolia*) with three-lobed leaves is a natural hybrid between the two.

To the characters of *P. orientalis,* the London Plane has added an ability to thrive in the poorest soil—apparently resigned to growing between the flagstones of a street—and in smoke-laden atmosphere. The only reaction seems to be a more frequent shedding of the bark. See Plate 16.

If the plane is the sycamore of the Americans, the issue is further complicated in that the name sycamore is nearly always used in Britain for the Great Maple (*Acer pseudoplatanus*) which is another alien coming originally from central Europe and western Asia. In contrast to the native common or small maple (*Acer campestre*) which grows only to a small tree size usually less than 20 feet high, it may grow to a well formed tree of 70 feet. Both yield a compact fine grained and attractively figured wood. An allied species, the Norway maple (*A. platanoides*) was introduced into Britain in 1683 and grows to a height of 100 feet. Unfortunately the maples which give such incredible beauty to the Canadian scene in the fall, when whole trees remain in full leaf yet every leaf a translucent fiery red, behave differently under the climate of Britain. Spectacular autumn tints are rare or occur only in exceptional seasons.

The lime or linden (*Tilia*) has already been mentioned but it would seem that only one of the species or sub-species can be regarded as a home product. This is the small-leaved *T. cordata* whereas the large-leaved lime (*T. platyphylla* or *T. grandifolia*) is planted, not wild. This last is native in Europe south of Denmark. The so called Common Lime (*Tilia vulgaris* or *T. intermedia*) is a hybrid between the two. Thus the trees which form those famous avenues such as that of Trinity College, Cambridge, are actually aliens. They, in contrast to the native variety, reach a very large size. All the limes have beautiful translucent foliage in spring which for a short time may be a clear golden yellow in autumn. The flowers are very rich in nectar and attract many bees : but it is honey-dew derived from the leaves by aphides which makes the leaves sticky and is a great disadvantage in smoky towns. The inner bark of the lime yields bast, used from earliest times for tying garlands : the soft white wood is valued for making such special items as architects' models and light bowls. The American lime (*T. americana*) or basswood yields the American white wood, also light and soft. This tree was successfully grown by Philip Miller at Chelsea in 1752.

An excellent example of a lime avenue is shown in Plate 15.

Like the sweet chestnut, the slow growing walnut has a nut greatly valued since very early times. Its original home is unknown: the designation Persian walnut expresses the belief that it originated in western Asia. It was certainly known to and used by the Romans. Was it introduced into Britain by them? This is possible, yet the earliest date definitely recorded in England is 1562—a difference of some fifteen centuries.

Another introduction credited to the 16th century is the holm oak (*Quercus ilex*) a native of Mediterranean lands (Plate XIXb). It is often called the evergreen oak because its leaves last for two years and this favourite shade tree is always green. The pale young leaves soon become dark, pointed, and often spiny along the edge, hence the Latin specific name *ilex* (holly). Somewhat later, probably in the first half of the 18th century, the Turkey oak (*Q. cerris*) was introduced from Turkey. It grows to a more regular shape and size than our English oaks and has deeply lobed leaves. From eastern North America come the Red Oaks but their foliage rarely turns to quite the same deep rich red which emblazons the American fall.

The poplars form an interesting and varied group of trees. Standing by itself is the aspen (*Populus tremula*) widely distributed though not a dominant in the great spruce-pine forests of northern Europe. There it reaches to within the Arctic Circle, beyond the limit of the conifers to the margins of the tundra itself, where it is associated with the lowly dwarf birch (*Betula nana*) which grows only a few inches high. It is not surprising to find the aspen often abundant in Scotland and recorded up to 1,600 feet above sea level. Another British native is the grey poplar (*Populus canescens*) with slightly hairy undersides to its leaves and which is very closely allied to the white poplar (*Populus alba*), a handsome tree, sometimes growing to 70 feet high, and which yields a useful soft white wood. The white poplar was probably introduced from Holland. The other common poplar, the black poplar (*P. nigra*) is probably not a native either. It comes from central and southern Europe and has a yellowish wood. It is widely planted in Britain. Special mention must be made of the tall slender quick growing Lombardy poplar, a variety of the last (*P. nigra* var. *italica*). It is probably native to Persia but has long been widely planted in Europe where, as in Italy, France and Holland, its clean vertical lines form a desirable contrast to the long

Plate 13. LOMBARDY POPLAR
A magnificent example of this Eighteenth-Century introduction (*p. 176*)

Plate 14. BLACK POPLAR IN WINTER
This fast-growing and graceful tree was probably introduced from continental
Europe in the middle ages   (*p. 176*)

horizontal lines of flat cultivated land, especially when planted as an avenue lining a main road. It is said to have been introduced into Britain soon after 1750 or perhaps even earlier—some claim by Lord Rochford in 1758. There are many American species of poplar which are occasionally seen in Britain. Americans often become nostalgic over the cotton-woods (*P. heterophylla*) very widespread, handsome and homely yet a valuable source of soft light tough wood. Other American cottonwoods are grown in Britain, notably the Oregon black balsam poplar (*P. trichocarpa*). However, most poplars now grown in Britain are hybrids between *P. nigra* and *P. deltoides*. How familiar poplars have become in the British rural scene is evident from Plates 13 and 14 and XIXa.

It would seem that willows (*Salix*) have always been a feature of damp localities, notably stream banks, in Britain. At least 12 or 15 species apart from varieties are recognized by botanists : some have listed as many as 80. *Salix alba*, the white willow, and *S. caprea*, the goat willow, are associated with Palm Sunday when branches with the male and the female catkins are gathered for decoration of church and home. *Salix alba* var. *caerulea* is the cricket-bat willow, famed source of wood for the making of cricket-bats. The young shoots of several species are used in basket making—either with the bark or from which it has been stripped and the canes bleached. Willows used in this way are known as osiers, and may be specially planted as osier beds. Naturally many willows have been introduced and garden varieties evolved in nurseries. It is said that the weeping willow was brought to Britain about the middle of the 18th century but the old assertion that "it was under the weeping-willow that the children of Israel mourned their Babylonian captivity" is probably pure phantasy.

The Common or Norway spruce (*Picea abies* or *P. excelsa*), Britain's favourite Christmas tree, may be regarded as one of the very common European forest trees accidentally excluded from the country after the Ice Age. It had previously been a native tree but had been driven out as the ice advanced. It was introduced or rather re-introduced at least as early as the beginning of the 16th century. It is well-named as characteristic of Norway but occurs practically throughout northern and central Europe. Its common associates in the Scandinavian forests is the Scots pine, the spruce predominating on the wetter situations, the pine on the drier or sandier. The use of the spruce as the Christmas tree is of Teutonic origin and in England dates only from about the eighteen-thirties, being popularized by Prince Albert in 1841. The ordinary

ML.—N

European larch (*Larix europaea*) was another accidental exclusion and was introduced in the early 17th century with the common silver fir (*Abies pectinata*) about the same time. The Corsican pine (*Pinus laricio*), first planted in 1759, is a Mediterranean species particularly common in Corsica but is regarded by some botanists simply as one variety of Black pine (*P. nigra*) of which another is the Austrian pine (var. *austriaca*) from the mountains of Austria and the Balkans. The Austrian pine has a more bushy habit than the native Scots pine and because it will withstand strong gales and poor soil is planted as a nurse tree on sand dunes, slag heaps and waste ground. But its wood is coarse, very full of knots and of so little value that the Corsican pine is more often used.

A fine specimen of a larch is shown in Plate 17 and cones of the Norway spruce and Atlas cedar are shown in Plate XXI.

The umbrella or stone pine (*P. pinea*) is another introduction from the Mediterranean said to have been made as early as 1548 (Johns). It is the tree inevitably associated with coastal scenery of the Mediterranean and familiar through the work of many artists to those who have never seen the Mediterranean for themselves. Opinions differ as to whether it should be called a beautiful tree : intriguing is perhaps a better adjective. Rather poor specimens are grown in south-west Britain for decorative, not commercial purposes. More common in the south-east near London and south-west is *Pinus radiata*, mentioned below.

The pinaster or Maritime pine (*P. pinaster*) is the pine which binds the sand dunes of the Landes in south-western France and there provides pit props and charcoal as well as resin and turpentine. It has become firmly established in a few parts of southern Britain notably the Isle of Purbeck.

The exploration and settlement of the New World brought to the knowledge of the peoples of the Old World a wide range of both beautiful and useful trees. Knowledge of food plants such as maize, birds such as the turkey, spread rapidly but it was much later before American trees were introduced to Europe. The slow wooden ships of the time were not well adapted for transport of young tree seedlings. One of the early introductions was the Weymouth, or White, pine (*Pinus strobus*, the source of American white pine) which was brought to England from eastern North America early in the eighteenth century by Lord Weymouth (1710) ; nearly all the other familiar American importations did not take place until the nineteenth century.

There is indeed a very interesting connection between the opening up of the American west and the introduction of North American trees into

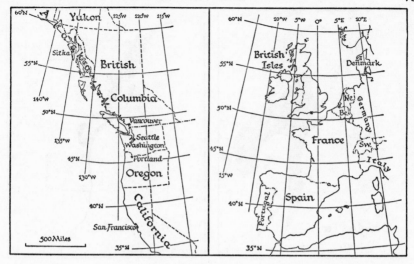

FIG. 30

Outline maps of western North America and western Europe on the same scale and showing same latitudes

Britain. The Erie Ship Canal established water connection between New York and the Great Lakes in 1823 and marked the beginning of a period of relatively easy movement between the Atlantic seaboard and the heart of the United States where only slow lumbering wagons had penetrated earlier. There followed a quarter of a century of expansion and movement ever westwards : to the Mississippi, beyond to the Rocky Mountains and the forests of the Pacific margins. For vast numbers California came into existence with the Gold Rush of 1849. The Scottish plant-hunter, David Douglas, who perished in 1834 in Hawaii—then known as the Sandwich Islands—was responsible for at least four noteworthy American introductions. In 1827—28 he brought to England for the first time seeds of one of the noblest of great trees, that magnificent timber producer of British Columbia and the Pacific North-West, the Douglas Fir. Surely no tree has ever been so insulted by its inadequate botanical designation, *Pseudotsuga taxifolia*—the false hemlock with leaves like a yew. Those who know the British Columbia coastlands and believe them to be scenically unsurpassed in the world, with snow-capped peaks towering over tree-clad slopes, plunging to great depths in blue-watered fiords, will appreciate how much is owed to this magnificent tree, often towering unbranched to 250 feet, and how

FIG. 31 (*left*) and FIG. 32 (*right*)
The Distribution of Douglas Fir (*Pseudotsuga taxifolia*) and Sitka Spruce
(*Picea sitchensis*) in North America

The Douglas Fir comes from latitudes comparable with southern England and
Wales, whereas the Sitka Spruce thrives in latitudes extending beyond the
northern limits of Scotland

much it may yet contribute to the development of new beauty in Britain.
When young the bark is silvery and almost smooth, but with age
becomes deep red in colour and deeply fissured. On a wood fire it burns
like good coal.

Four years later, in 1831, Douglas brought to Britain the now well-
known Sitka spruce (*Picea sitchensis*) from even farther afield in the
American north-west. The next year he introduced the well-named
Giant Silver Fir or Grand Fir (*Abies grandis*) and in 1833 the Monterey
Pine (*Pinus radiata*) which grows well in the south and south-west of
England and is distinguished by its bright green foliage.

The Monterey Pine should not be confused with the Monterey
Cypress (*Cupressus macrocarpa*), now such a familiar and favourite hedging shrub in southern Britain. *Cupressus macrocarpa* was once widespread
in America, but failed to adapt itself to the climatic fluctuations of the
later ice age and survived only in the tiny peninsula of Monterey in
California from which it takes its name. There one finds gnarled old
specimens apparently fighting a losing battle for the continued existence

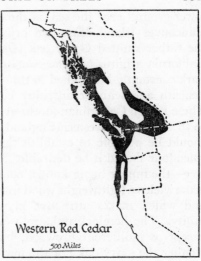

FIG. 33 *(left)* and 34 *(right)*
The Distribution of Western Hemlock (*Tsuga heterophylla*) and Western Red
Cedar (*Thuya plicata*) in North America
Both come from latitudes comparable with the British Isles

of the species in face of Pacific winds and fogs. It is the extremely rapid
growth—a tall, thick, evergreen hedge in three or four years—which
renders *macrocarpa* so popular in new housing estates in Britain, but it is a
fickle friend and liable to die off in ugly brown patches if trimmed as a
hedge. Every half-dozen years or so a severe frosty spell will kill off a
large proportion, though after a number of years mature trees seem to be
completely hardy at least in southern Britain, and gnarled old specimens resemble those in their native Californian home. *Cupressus macrocarpa* was brought to Britain in 1838 ; the almost equally well-known
Lawson's Cypress (*Cupressus (Chamaecyparis) lawsoniana*), the name of
which commemorates an Edinburgh nurseryman, not until 1854.
Though grown as a hedge shrub, it makes an excellent tree. Another
American introduction of about the same period—actually in 1851—
was Western Hemlock (*Tsuga heterophylla*). Lawson's Cypress is
illustrated in Plate 18 and the Monterey Cypress in Plate XX.

In the great forests which clothe much of Vancouver Island and the
coastal ranges of British Columbia three conifers share dominance. They
are that great timber tree the Douglas Fir, the Western Red Cedar
(*Thuya plicata*) and the Western Hemlock (*Tsuga heterophylla*). On the

lower ground near the sea on the drier Gulf Islands and neighbouring
Vancouver Island coast two broad-leaved trees become abundant—
the rather stunted Garry oak (*Quercus garryana*) and the Madrona or
California arbutus (*Arbutus menziesii*) a most attractive tree with a red
bark constantly being shed in thin sheets to expose a bright green layer
beneath and with broad shiny laurel-like leaves shed at any season.
These British Columbian forests grow under climatic conditions so close
to those of south-western England, Wales and western Scotland that it
should be possible to establish them in Britain in all their matchless
splendour should it be desirable. The Douglas Fir is a welcome timber
tree—the timber being known commercially as Oregon pine. The red
cedar yields a lightweight wood free from knots ideal for roofing shingles
and which makes attractive plywood for panelling, but of limited
utility. The hemlock unhappily was long regarded as of little value
except as indifferent firewood, but is now a valued pulpwood tree.
Neither the garry oak nor the madrona have special claims to be considered as timber trees. There is a solitary specimen of the *Arbutus,* old
and carefully guarded by iron railings, in Kew Gardens : whether it
would thrive as a curiously fascinating rather than ornamental tree in
the mild south-west of Britain has probably never been put to the test.
The hemlock however has long been valued as an ornamental tree in
British parks : it has indeed all the virtues to commend it for such use.
It grows to a graceful pyramidal form, the top of the main stem bowing
over gracefully to the breeze, its branches sweeping downward then
curving up. The young flattened needles a half to three-quarters of an
inch long are pale green, becoming darker with age but retaining a
silvery underside. The numerous pale green young cones on the fruiting
branches are an added attraction. How unfortunate the timber is such
that the foresters cannot wax enthusiastic! When Douglas Firs are
mature the foliage where exposed to wind in the upper part becomes
sparse, some would say ragged, and they are most effective when their
slender vertical lines stand out against the evening silver of the fiords in
their native home, backed by the purple of the mountains. The
hemlock on the other hand retains its graceful shape and foliage with
maturity and hence its reputation as an ornamental tree.

These Douglas Fir-hemlock-cedar forests of the more sheltered parts
of the British Columbian coast give place northwards to almost pure
forests of Sitka spruce (*Picea sitchensis*). Nothing seems to deter this
hardy tree : it grows right down to the water's edge, sometimes
drenched with Pacific ocean spray, and towers magnificently to 200 feet

Plate XXIa. (left) THE FOLIAGE AND CONES OF THE NORWAY SPRUCE, FOREST OF DEAN, OCTOBER (p. 177) (John Markham)

b. (right) THE FOLIAGE AND CONE OF THE MOUNT ATLAS CEDAR, MONMOUTH, OCTOBER (p. 185) (John Markham)

or more. Its foliage is stiffer than that of the Norway spruce and it has a bluish tinge. The whitish wood is very strong yet light and so particularly valuable in joinery as well as making good plywood and pulpwood. It is small wonder that it is fast becoming Britain's dominant forest tree in the wetter western lands where other species give up in despair (see Plate 19).

It is from farther south, in the great forests of Washington, Oregon and northern California that we derive the Noble Fir, *Abies nobilis,* classed as one of the "minor species" now being planted by the Forestry Commission, but well known in parks. Truly a noble tree, it is to be hoped it will succeed as a timber tree in the new Britain. More important economically from parts of these same forests is the Lodgepole Pine (*Pinus contorta*), so called because the Indians used its long slender poles for their tepees. The misleading specific name results from the tree first being described from contorted coastal specimens.

FIG. 35
The Distribution of Lodgepole Pine (*Pinus contorta*) in North America It comes from latitudes comparable with those of the British Isles. On the whole it is dominant away from the sea

The California redwood is a magnificent tree rightly associated with the coastal belt of California north of San Francisco and through which now runs the scenic Redwood Highway. There a number of groves have been saved from the lumberman's axe and the trees ranging in age from 400 to 2,000 years for mature specimens tower unbranched to heights of over 300 feet. One at least has been measured and found to be 364 feet high. The Redwood is a noble and a beautiful tree, its deep red bark contrasting with its deep green foliage. Known botanically as *Sequoia sempervirens* it was brought to England in 1843 or 1846 and is still sometimes listed in nurserymen's catalogues as *Taxodium,* its name before the genus *Sequoia* was introduced by Endlicher in 1847 (see Plate XXIV).

Plate XXII. THE CHILE PINE OR MONKEY PUZZLE (*Araucaria*)
One of the few introductions from South America. (*Eric Hosking*) (see page 186)

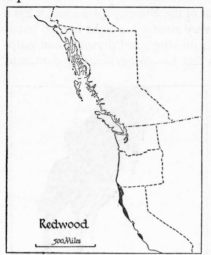

FIG. 36 (*left*) *and* FIG. 37 (*right*)
The Distribution of Redwood (*Sequoia sempervirens*) and the Big Tree
(*Sequoia gigantea*) in North America
Both are from latitudes comparable with Spain and Portugal—not with the
British Isles

The Redwood could claim to be the world's biggest tree if it were not for the other species of the same genus *Sequoia gigantea*, the Big Tree of California, found growing mainly between 4,000 and 5,000 feet above sea level on the Californian Cordillera. Americans of today have a great pride in the giant sequoia which they claim as the largest and oldest living things on earth. Many reach over 300 feet in height and may be over 4,000 years old ; some with timber and foliage weigh over 2,150 tons. Though archways have been cut through redwoods, *Sequoia gigantea* is the original big tree of the horse and buggy days—driving at full speed through the natural arch. Some of the finest specimens have now been preserved in national and state parks, especially in the Sequoia National Park. The tree was brought to Britain by William Lobb in 1853 and named *Wellingtonia* (in honour of the Duke of Wellington who had died the previous year) by Lindley. In Britain this name has persisted but, not unnaturally, did not find favour in America. After some attempts to use *Washingtonia,* the affinities of the tree are recognised by the general use of *Sequoia* in both popular and technical circles. Though not greatly valued in Britain as a timber tree, it has become one of the best known conifers in parks. Some admire its tall conical form, others consider its beauty marred by a certain poverty of foliage (see Plate 3b and Plate XXIII).

The two Sequoias are of very great interest from another point of view. Botanically they are survivals from mid-Miocene times—many millions of years ago. Both species are found fossil in Miocene deposits of Oregon at which time they had a circumpolar range. A still earlier form of these living fossils has been found in recent years in China (*Metasequoia*). It has been introduced into botanical gardens and several plantations exist in Britain.

There is a point of very great geographical interest about these successful introductions from America to Britain. They are from the Pacific North-West of the United States and Canada where climatic conditions most closely resemble those in the British Isles. In the eastern parts of North America with very hard winters and hot summers are many trees which have not proved successful in Britain. Similarly the long-leaf pine (*Pinus palustris*), yielding a magnificent timber in the warm south-eastern states, does not flourish in Britain but *Taxodium disticum*, the Swamp Cypress, from the same region grows well.

Western North America was not the only source of trees introduced into Britain about a century ago. Some of the magnificent forests of the Indian Himalayas inspired even the British Government of the day. The Bhutan or Blue pine had reached Britain in 1823 and the Deodar or Himalayan cedar (*Cedrus deodara*) in 1831. It was the rapid growth of the seedlings of the deodar and the high quality of the Indian timber which led the British Government to import quanitities of seed so that deodar timber produced in Britain might provide a substitute for oak in naval shipbuilding. As an ornamental tree the deodar grows well in Britain but some users consider the timber inferior to that of the Atlas cedar introduced from North Africa about 1845, or the wonderful cedars of Lebanon (*Cedrus libani*) of biblical fame (now the symbol of the Lebanese Republic) introduced into Britian at least as early as the seventeenth century. Few trees have inspired poets and song writers through the ages as has the cedar of Lebanon. To the writers of the Psalms and the author of the Song of Solomon it was a symbol of longevity, power and beauty: so too for Shakespeare and Laurence Binyon. Today one searches with care for the few remaining groves on the slopes of the Lebanese mountains. In Britain many of the fine old specimens associated not unnaturally with Bishops' palaces (*cf.*Plate 3a) have gone.

The Far East yielded its quota of trees introduced into Britain. In 1842 came the beautiful *Cryptomeria japonica* from Japan, followed in

1853 by the giant thuja (*Thujopsis dolabrata*) also from Japan—another of William Lobb's introductions. Most of these introductions provided ornamental specimen trees in parks and gardens but in 1861 we were destined to receive from Japan a timber tree now of much significance—the Japanese larch (*Larix leptolepis*). It is not such a tall or strong tree as the European larch, but it grows more quickly and under harder conditions. The Japanese thuja is now commonly used as—one might say demoted to—a hedge plant. The thuya planted by our Forestry Commission is however the American Western Red Cedar (*T. plicata*) already discussed.

South America has provided us with at least one notorious exotic beloved of Victorian gardeners—the monkey puzzle or *Araucaria imbricata* (Plate XXII). It is a slow growing tree of the Cordillera of Chile where it grows to a height of 150 feet. It was introduced into England in 1796 and is hardy in nearly all parts of the country, though liable to damage by severe frosts. Its near relative, *A. brasiliana* (introduced in 1819) proved to be non-hardy.

No attempt has been made in this chapter to be exhaustive. Its object has been primarily to call attention to those many trees introduced into this country which have either become so closely naturalized as to be regarded as natives or which, by extensive use in park and forest, play a considerable part in the landscape of the countryside. It has not been possible to consider the many trees which are essentially garden dwellers—one thinks at once in this connection of the laburnum introduced from France some centuries ago, the flowering cherries and Judas trees—but only those which range beyond the garden to become trees of the countryside.

It must be emphasized that it is still the native trees which dominate the landscape over the greater part of rural Britain, especially the lowlands. The innumerable little woodlands are mainly of native oaks, the hedgerow trees which are such a characteristic feature of our country are usually natives—notably elm.

Those who seek a full account of the native woodlands of Britain are referred to Sir Arthur Tansley's *British Islands and their Vegetation*. Many trees not mentioned in the foregoing chapter will be found described and illustrated by L. J. F. Brimble in his *Trees in Britain* or by H. L. Edlin in his *British Woodland Trees*.

In the next chapter we take up the question of the new forests now being created in Britain, and the trees planted in them destined to make a major change in our country landscape.

Eric Hosking

Plate XXIII.—*Sequoia gigantea* (Wellingtonia) in Coniston, Lake District

CHAPTER 14

TREES OF THE FOREST, 1919*

THE YEAR 1919 marked a turning point in the story of scenic evolution in Britain. As long ago as 1503 an Act of Parliament referring to England stated quite simply that the forests of the country had been "utterly destroyed". John Evelyn in his *Sylva : or a Discourse of Forest Trees*, published in 1664, says, "truly the waste and destruction of our woods has been so universal that I can conceive nothing else than universal plantation of all sorts of trees will supply and encounter the deficit". From time to time others voiced similar sentiments : the need for oak for ship-building led to a little attention being given to the management of such oak woodland as remained. In England some of the great landowners, especially in the latter part of the 18th century, started the work of planting but rather for the embellishment of their estates than with the production of timber in mind. It became fashionable for landowners to vie with one another in planting newly introduced trees and especially in arranging them in their parklands ; scattered so as to display the peculiar beauties of each, formally to aid the overall design of house and garden or in clumps to serve as distinctive landmarks. Scotland which was described by one author in the mid-eighteenth century as "a land unenclosed, hedgeless and treeless" shared in this change. The planting of trees was much encouraged by the establishment of the Highland Society in 1783 and there it was regarded as a long-term investment by many landowners.

The key to the problem is in the words long-term investment. The period required for trees to reach maturity or at least a size which will yield timber varies in Britain from about 60 to 200 years according to

*I am greatly indebted to Mr. H. L. Edlin, author of *Trees, Woods and Man* in this series, and to Professor Sir H. G. Champion for valued comments on this chapter.

Plate XXIV. REDWOODS IN MONTGOMERYSHIRE
These trees were planted in 1860 and were 93 years old when inspected by the Royal Forestry Society in September, 1953 (see page 183). (*The Times*).

climate, elevation and soil, and the species of tree. Though there are quick-growing poplars and willows, hardwoods have generally a longer period of rotation than conifers. An oak requires 120 years to reach maturity, a Scots pine perhaps about 80 years. These figures are of course rough approximations only but quite clearly a man does not plant for his own benefit but for that of his descendants. So long as great estates remained in family hands the planting of trees could be regarded as long-term investment. The introduction of, and great increase in, death duties or inheritance taxes virtually removed the incentive for such altruistic if necessary expenditure by landowners, despite some tax concessions on standing crops of timber.

Thus, as the nineteenth century went on, remaining forests and woodlands of any value for timber production shrank still further. Just as the expansion of British exports of manufactures to all the new countries of the world brought an automatic inflow of foodstuffs and led to the steady neglect of the land and the decline of home food production, so it also brought a comparable inflow of raw materials including timber and pit-props, pulp and paper and led to a complete neglect of forestry at home.

So Britain became the least wooded of all the countries of Europe with under five per cent. of its area under trees. In Ireland the proportion dropped to under one per cent. Even including scrubland of little or no economic value, cut-over land not replanted, and purely amenity woodland, the total for England and Wales in 1913 was only 5.0 per cent. and Scotland 5.6. Less than half of this was high forest or plantation.

Such old forests and woodlands as Britain retained (and still retains) owe their preservation to historical or accidental causes. In the first place there are the ancient Royal Forests originally for the use of the King and his friends for hunting. There are five such named in Domesday—the New Forest, Windsor Forest, Wychwood, Grovely and Wimborne. Other tracts were designated as Royal Forests later and similar lands were likewise preserved for hunting by the landed nobility. It will be remembered that the early use of the word "forest" applied to any area set aside primarily for sport whether occupied by trees or not. This old usage is still current in place or district names such as Dartmoor Forest, Exmoor Forest, Rossendale Forest, Bowland Forest and others. Similarly the 'Deer Forests' of Scotland are almost treeless. The Royal Forests mentioned in Domesday, however, included actual wooded areas and it is practically true to say that all the major

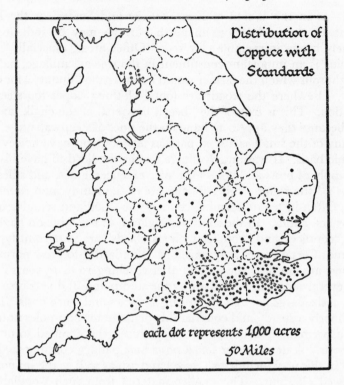

Distribution of Coppice with Standards

each dot represents 1,000 acres
50 Miles

FIG. 38
The Distribution of Coppice-with-standards in Britain

forests existing in England in 1913 owed their preservation to the sport-ing instincts of the King and the nobility. They included the New Forest Forest of Dean, Delamere Forest (Cheshire) and Sherwood Forest (Not-tinghamshire), whilst most of Windsor Forest had become Windsor Great Park. Other famous forests—Arden in Warwickshire and Rocking-ham in Northamptonshire—had almost disappeared though per-petuated by names on the Ordnance maps. In such cases the maps often show tiny patches remaining (as in many parts of the country) and the designation "fox-covert" is sufficient indication of their purpose.

Where forest and woodland have been allowed to remain it is usually an indication of poor land. Rarely did either the landowner or the farmer permit good productive land to remain thus when a more remunerative use could be found. Considerable patches of woodland in Lowland

Britain—as in the Weald of Kent, Surrey and Sussex, or the Blean in Kent, or Ashtead Woods in north Surrey—may be indicative of extremely heavy intractable clay soils. Other areas, especially where Scots pine is an important constituent in the tree assemblage, indicate very light soils which the farmer would consider too "hungry" for cultivation. Elsewhere the woods are found clothing slopes too steep for cultivation. This is true of the "beech hangers" of the chalk lands so called because they appear to hang on the sides of steep valleys.

Many of the small scattered patches of woodland have however an economic basis. It was an advantage for the landlord to have his own local supply of posts and poles for repair of fences, gates, and stiles and perhaps a little rough timber for the use of the handy man round the estate. Often in tenant-farming country the woodland remains in the landowners' hands. In the south of England, as the table on page 203 shows, "coppice" and "coppice with standards" are important types of woodland. Certain trees may be left as "standards" to grow normally: the others are cut near the ground after reaching 10 to 15 years in age when several shoots are given off by the stump and in due course each reaches a size large enough for cutting for poles and fence posts. There was formerly a steady and considerable demand for hop-poles until the use of a permanent structure of wire reduced the demand to one for main posts. A demand for fence posts and palings continues. Trees treated in this way include oak, hazel and ash as well as the supposedly introduced chestnut. The smaller material from such woodlands is useful as bean and pea-sticks. In some areas woodlands are cut regularly for firewood. Little saw mills, some driven by water, still survive in many parts of the country and in the years following the Second World War when coal to the domestic consumer was rationed a large temporary trade in logs cut to about nine inches long developed.

Much of the deforestation of Britain is traceable to the early iron smelting industry when charcoal was the fuel used. The once-extensive iron industry of the Weald disappeared with the exhaustion of the wood-charcoal. A little charcoal is still made in parts of the country, mainly in the Southern counties.

In a country such as Britain where many parts are subjected to strong and often persistent winds woodland has an important protective function. Some of the narrow "shaws" between fields are remnants of a former woodland cover—this is the case in the Weald. Elsewhere shelter belts have been planted, usually at right angles to the prevailing wind or winds when the seriousness of complete clearance has been recognized.

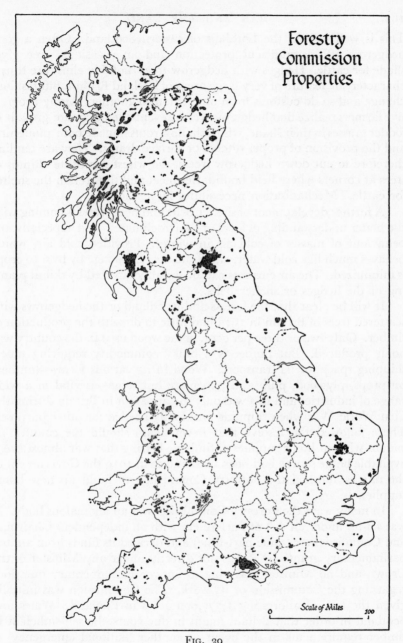

FIG. 39

The Distribution of Forestry Commission Properties in Britain, 1954

This is well seen in the Lothians of eastern Scotland. Often a good hedgerow affords sufficient protection and an occasional tree gives shade for cattle. Hedges with hedgerow trees (notably elms) are thus a characteristic feature of very large areas of lowland Britain. But fashions change and so do customs from one region to another. Many progressive farmers realize that hedgerow trees prevent the adequate growth of fodder grasses in their shade, whilst shallow roots impede both ploughing and the provision of proper ditches or surface drains. They are tending therefore to cut down hedgerow trees: the wisest substitute clumps of trees at corners where field boundaries meet and thus retain the shelter for cattle. Mechanisation necessitates larger fields.

A further development of shelter belts and hedgerows is coming with the better understanding of local or "microclimates" and especially the behaviour of masses of cold air in winter. Provided cold air, which behaves much like cold water, can drain away, damage by frost to crops is minimized. The air currents can be suitably directed by skilful planting of the hedges or shelter belts.

It will be clear that the forests and woodland or the hedgerows with scattered trees of Britain in 1913 had little to do with the production of timber. Only two or three per cent. of the wood used in the country was home produced. But timber is a bulky commodity requiring much shipping space for its transport. Wood in its various forms—lumber, pit-props, plywood, pulp and paper are but some—is vital in a wide range of industries and this was only too apparent in Britain during the First World War when shipping was badly needed for other purposes. There followed the resolve that never again should the country be caught without home supplies of timber. During that war almost every available home source had been used and so in 1919 the Government of the day established the Forestry Commission to build up new home supplies.

In many ways the Forestry Commission was an anomalous body. It was set up under an Act of Parliament with an independent Chairman and Commissioners (some part-time) and derived its funds from annual parliamentary grants yet it was not directly under any Minister of the Crown and no Minister was able to answer Parliamentary questions regarding the Commission or its work. The Commission was initially given the task of afforesting 1,777,000 acres in England, Wales and Scotland in order that Britain might in due course be self-sufficient in timber requirements in the event of another 'national emergency' of three years' duration. Actually the 'emergency' in the shape of the

C. W. Bradley

Plate 15a. LIME AVENUE IN SUMMER

C. W. Bradley

b. LIME AVENUE IN WINTER
The large-leaved lime was introduced in the Sixteenth Century. This avenue is at Aston Park, Cheshire

Plate 16. PLANE TREES IN BERKELEY SQUARE

Second World War came long before the work of the Commission had reached the necessarily advanced state.

In its early years the Forestry Commission experienced many difficulties. Its Australian-born Chairman, Sir Roy Robinson (later Lord Robinson), though he made enemies by his forthright methods, persisted and triumphed where many a lesser man would have quailed and failed. The Commission's grant was subjected to cuts by Parliament and petty economies which completely ignored the long-term nature of forestry work. The Commission was compelled to acquire land in the open market or in open competition and was limited in the price per acre it was permitted to pay. Not infrequently a whole estate had to be purchased although only a relatively small acreage was plantable. This resulted in hill-sheep land being taken at the very time when the Ministry of Agriculture was under obligation to help hill-sheep farmers and there was thus a conflict between a Government department and a Government sponsored Commission. Between 80 and 90 per cent. of the timber used in Britain is softwood and, quite rightly, the Commission paid first attention to these timbers. The Commission came in for severe and loudly voiced criticism both because it favoured the exotic conifers and not the native hardwood trees but even more so because of what country-lovers saw to be happening as a result of the Commission's activities. All planting involves giving the proper spacing to each young tree and this results in a certain regularity of arrangement in rows and blocks somewhat at variance with the natural contours of the landscape. Critics saw moor-covered hillsides which tended to assume an aesthetic value they had never before possessed being replaced by a coniferous rash for which their language failed to find adequate adjectives. At that time professional foresters considered it easier to manage forests which were in regular blocks, fire hazards were best met by long straight rides, so it is quite true that alien rectangular patterns began to develop over parts of the more rugged countryside and pure stands of alien trees appeared where native mixed hardwoods once had been.

Actually both features of the Forestry Commission's early work which gave most offence are probably bad forestry. The natural upper limit of tree growth follows a sinuous not a straight line : it is low over bare windswept shoulders and creeps higher up sheltered gulleys. There are natural routeways into hill-land along which eventually the timber is most easily withdrawn : thus the hard straight line is nearly always wrong though it may represent economy in fencing. In natural regeneration of forest pioneer species come in and colonize the land and act as

ML.—O

nurse crops to the trees which will ultimately establish themselves as the dominants in the climax vegetation. In nature pure stands are rare except in patches specially favoured in natural conditions such as micro-climate, elevation, aspect and soil for one particular species : elsewhere two or more codominants fight it out between them and flourish better than any one would do alone. Stands of a single species are peculiarly liable to epidemic diseases. Robinson himself, as the work of the Commission proceeded, recognised that the foresters' work should be to speed up the natural process of plant succession : to use larch as a nursemaid to beech or certain hardwood species as nursemaids to a mixture of conifers all of which were of commercially desirable species. Just as in the great boreal forests of Scandinavia and Finland, Norway spruce and Scots pine are codominants, the first developing best in damper situations the second in drier but both benefiting from an intermingling of birch and occasional aspen, so the new forests of Britain are likely to flourish best if there is an essential element of admixture of species. If this is so, the second main aesthetic objection to the re-forestation of Britain disappears.

The argument that forests are not 'natural' to the landscape of Britain we have seen to be the reverse of true. Afforestation gives us the opportunity of recreating beauty we have lost for centuries as well as of adding to the wealth of our country a much needed natural resource. When so many of the features we admire today in our scenery, so many loved trees, are really only aliens of long standing, it seems nonsense to talk about the evils of introducing 'exotics'.

From its establishment in 1919 the work of the Forestry Commission, despite setbacks, went on steadily but the "emergency" for which it was set up came long before its plantings were ready for use. The oldest of its plantations were only 20 years old in 1939 on the outbreak of the Second World War, although by that time some were yielding useful thinnings—with some timber big enough for use as pitprops. In 1939 less than 4 per cent. of the home consumption of timber in Britain was home-produced. This excludes wood-pulp and paper. The Second World War brought a still further savage depletion of resources and left the country in a sorry state so far as useful timber was concerned though many young plantations were in good shape.

In 1945-46 the Government removed some of the anomalies in the status of the Forestry Commission by making it answerable to Parliament through the Minister of Agriculture and Fisheries and the Secretary of State for Scotland, acting jointly in matters of national

Plate XXVa.—Kirroughtree Forest, Kirkcudbrightshire
The Forest nursery with plantations of Corsican pine beyond
b.—Queen Elizabeth Forest, near Buriton, Hampshire
An example of hardwood planting by the Forestry Commission on the steep scarp slope of the chalk downs. The crop is beech with larch " nurses " to help it through the early stages

policy. The Commission did not become part of the Ministry of Agriculture nor of the Department of Agriculture for Scotland, but the new arrangement meant a careful integration of its work with agricultural interests. Estates comprising hill-land which fall to the State in lieu of death duties are to be managed jointly in the interests of food production and forestry : unplantable hill land is to be so made accessible as to serve as hill-sheep pastures.

Whilst the war was still in progress a greatly expanded programme for the Forestry Commission was set out in a Government White Paper*. This declaration of policy laid down that 3,000,000 acres were to be afforested ; 2,000,000 acres to be replanted, whilst inducements were offered to private owners to 'dedicate' woodlands which the Commission would then require to be managed according to an approved plan, the Commission paying grants for planting and maintenance. When it is remembered that the *total* area of all high forest and plantations in 1913 was only 1,416,890 acres (England, Wales and Scotland) the 5,000,000 acres mentioned is a big figure. It is nine per cent. of the total area of the country. Indeed if the whole programme is completed the area of forest and woodland may rival that apparently destined to be covered by bricks and mortar†.

It is clearly of the greatest interest to discover what is this new landscape being created by the Forestry Commission. Fortunately in choosing trees which will both flourish in Britain and yield timber of value the Commission can draw upon the experience of those private landowners who planted for the embellishment of their parks, for pleasure rather than for profit, the various species from time to time introduced into this country. As we have seen in the last chapter nearly all the introductions were effected by, or soon after, the middle of last century so that the results of a hundred years of growth can be observed.

How completely the new forests will differ from the old is apparent if one looks at a popular work of a hundred years ago such as *English Forests and Forest Trees* published anonymously by Ingram, Cooke and Co. in London in 1853. There the author lists and describes the oak,

Post-War Forest Policy, H.M.S.O. Cmd. 6447, 1943.
†In 1931-34 the Land Utilisation Survey of Britain recorded 1,719,900 acres occupied by houses with gardens and 1,399,200 acres agriculturally unproductive, mainly built over, making a total of 3,119,100 acres. At the same time land was being lost to housing and industry at the rate of 50,000-60,000 acres a year.

Plate XXVI. THE SPRUCE WOODS AT GLEN BRANTER FOREST, ARGYLL
A National Forest Park. (*Forestry Commission*)

poplar, willow, ash, larch, lime, chestnut, elm, birch, pine, sycamore, walnut, beech, hazel, aspen, alder, plane, maple, hawthorn, holly, yew and hornbeam. By 1946-47 the dominant species in high forest had become, in order, oak, pine, sitka spruce, beech, Norway spruce, ash, birch, sycamore, European larch, Japanese larch, Corsican pine, Douglas fir, elm and Spanish chestnut.

For recent years some exact details are available—some fairly complete figures of acreage for 1913, a Forestry Commission Census mainly relating to 1924, one for 1938, and a thorough one in 1946-47. Taking other estimates into consideration the following table gives the official figures for total woodland area 1871 to 1962.

	England acres	Scotland acres	Wales acres	Great Britain acres
1871	1,314,316	734,530	126,625	2,175,471
1887	1,518,312	874,850	167,573	2,560,744
1895	1,665,741	878,765	181,610	2,726,116
1905	1,683,324	868,409	216,510	2,768,243
1913-14	1,667,574	852,120	216,494	2,736,188
1924	1,630,987	1,074,224	253,461	2,958,672
1938-39	1,809,800	1,076,300	315,000	3,201,100
1947	1,865,046	1,266,838	316,478	3,448,362
1962	2,102,000	1,637,000	466,000	4,205,000

The total area under woodland is thus about 4½ million acres or 7.5 per cent. of the total land surface of the country or 0.08 acres per head of population. Both these figures are lower than for any other country of significance in Europe and compare with the average for Europe of about 30 per cent. and 0.8 acres per head.

It is primarily the Forestry Commission which is changing the scenery of forest Britain and so it is interesting to note the species included in planting up to 30 September, 1962:

Sitka spruce	.. ..	420,000 acres	30 per cent.
Scots pine	.. ..	250,000 „	18 „
Norway spruce	..	172,000 „	12 „
Japanese larch	..	123,000 „	9 „
Lodgepole pine	..	80,000 „	6 „
Corsican pine	.. ..	72,000 „	5 „
Douglas fir	..	62,000 „	4 „
European larch	..	48,000 „	3 „
Other conifers	..	34,000 „	3 „
Oak	..	56,000 „	4 „
Beech	..	51,000 „	4 „
Other Hardwoods	..	31,000 „	2 „
Total	1,399,000		

It is useful to continue this table by figures of planting (1,156,665,000 seedlings) in the years 1953-62 by the Forestry Commission:

			Plants Used	Per cent.
Scots pine	..	..	205,675,000	18
Corsican pine	..	..	60,153,000	5
Lodgepole pine	..	..	140,095,000	12
Larches	..	..	143,835,000	13
Douglas fir	..	..	50,627,000	4
Norway spruce	..	..	012,084,000	9
Sitka spruce	..	..	298,037,000	26
Other Conifers	..	..	73,289,000	6
Oak	..	..	..	23,716,000	2
Beech	..	..	..	43,041,000	4
Other Broadleaved	..		16,113,000	1

I am greatly indebted to G. B. Ryle, Deputy Director of the Forestry Commission for this additional information. To the end of 1962 the Commission had planted 3,000,000,000 trees—seedlings and young transplants—in the proportion among species corresponding very closely with the period 1953-62 given above. By numbers planted conifers were 93 per cent., broad-leaved 7 per cent.; exotics 75 per cent., native species 25 per cent. (made up of Scots Pine 18 per cent. and broad-leaved 7). As the broad-leaved crops are usually started in a mixture with conifers (see Plate XXVb) to be removed later the broad-leaved plantings represent 10 per cent. by intended area. Commonly about 2,000 young trees per acre are planted and by the end of September 1963 the Commission had planted 1,860,000 acres.

The species have each their favoured habitats and regions in Britain. The Sitka spruce plantations are essentially a development for which the Forestry Commission is responsible. The tree was previously little known and little grown in Britain: now it leads in the Commission's newly planted areas and the near future will see it as Britain's dominant forest tree. In its native area it withstands the winds and rains as well as salt spray from the Pacific which sweep over the panhandle of Alaska and the coast of British Columbia: similarly it can withstand perhaps better than any other tree the likewise vigorous conditions of western Scotland. The two spruces, Sitka and Norway, are the trees of the wetter side of Britain—in Argyll, Kirkcudbright; in Cumberland and the Kielder forests of Northumberland and in most of the Welsh counties. In each case Sitka spruce has been used in afforesting the poorer and higher-lying types of wet grass moors with shallow or medium peat soils, whereas the Norway spruce has been planted where the land is lower, less exposed and the soils less infertile or less acid. In the eastern and lowland counties spruce usually marks patches of wet clay soil : in eastern Scotland the two spruces are of about equal importance, in the North Riding Sitka predominates but the smaller patches in eastern England are mainly Norway.

Scots pine was the tree of the native Caledonian forests of which remnants survive in the central Highlands, but there is no mistaking its preference for drier climates and drier soils. Heather moorland not on deep peat is usually indicative of conditions favouring the planting of Scots pine in contrast to wet grass moor where only spruce will grow and there has been extensive planting in all the eastern counties of Scotland as well as on the light soils of Breckland and other parts of lowland England. The Corsican pine from southern Europe has been found resistant to industrial smoke and fumes and has been planted extensively on the poor Bunter sandstone soils of Nottinghamshire and Staffordshire. It has also been used on the light soils of Breckland, in north Yorkshire and elsewhere but its use in Scotland and Wales has been almost restricted to the fixation of coastal sand dunes. On a small scale the American Lodgepole pine has been used on difficult moorland sites with infertile soils and considerable exposure in Scotland, northern England and North Wales.

The larches are more widely distributed than the pines. The European larch has long been planted in small patches because it is valued for general estate and farm use. Japanese larch can be grown well with higher rainfall and poorer soils. Hybrid larch is still most used in Perthshire where it originated.

Just as in western Canada Douglas fir, source of the commercial 'Oregon pine', is indicative of better conditions than those where Sitka spruce flourishes, so in Britain it is a tree of rather better lands. It is used particularly for the conversion of oak scrub of little value to high forest by underplanting and hence its importance in Inverness, Perth, Argyll, Kirkcudbright, Caernarvon, Merioneth, Radnor, Somerset and Devon.

Amongst the trees noted above Lodgepole pine (3,260 acres in 1946-47) and Hybrid larch (6,126 acres) are both called 'minor species' in the Forestry Commission's report. Others include tsuga or hemlock (1,076 acres—mainly in Scotland and Wales), thuya (359, mainly in England), and Lawson's cypress (335 mainly in England). The North American Grand fir (628 acres) and Noble fir (557) found mainly in Scotland and Wales are superseding the European silver fir (479 acres).

John Markham

Plate 17. THE EUROPEAN LARCH
A fine example of a tree introduced from continental Europe (*p. 178*)

Plate 18. LAWSON'S CYPRESS
Commonly used as a quick growing hedge shrub in gardens, this Nineteenth-
Century introduction from western America grows to a fine tree (*p. 181*)

Other conifers occupying together 2,694 acres include Austrian pine, maritime pine (locally naturalized in Dorset), deodar, Californian redwood, and yew.

The purpose of this chapter is to stress the spread, especially under the aegis of the Forestry Commission, of introduced forest trees. The table on page 196 is sufficient indication of the continued importance of hardwood species most of which, in contrast to the conifers, are native. The only two non-native broadleaved trees which play a considerable role in forest lands are the sycamore and sweet chestnut, both early introductions attributed to the Romans but which may, as mentioned above, prove to be natives.

It is interesting to note that the new forests of Britain illustrate a very significant principle. Where native species have been supplemented by or replaced by introduced species, the latter are, with few exceptions, either from neighbouring parts of Europe (Norway spruce and European larch) or from that part of the New World—British Columbia and north-western United States—where climatic conditions most closely resemble those of Britain. That climatically homologous area has given us Sitka spruce, Douglas fir, hemlock, red cedar, lodgepole pine, Grand fir, Noble fir, Monterey cypress and Lawson's cypress. Few successful trees come from climatic regions widely different from our own.

Norwegian experience with conifers from western North America would seem to indicate that best results are obtained when the trees are planted under climatic conditions slightly better—a few degrees higher in average temperature for example—than in the home environment. The success of Sitka spruce in Britain, under somewhat less rigorous conditions than in British Columbia, would seem to confirm this general conclusion.

There are still many unsolved or only partly solved problems of forestry affecting the planting programme in Britain and, consequently, the future of the landscape. One is the problem of local "races" or sub-species, which seem to be minutely adjusted to microclimatological conditions. Trees take a long time to grow and it may be anything up to 50 or more years before mistakes become apparent. It is accordingly most important to profit by the experience of other countries. Conditions in south-western Norway closely resemble those of western Scotland. There as in Scotland the Norway spruce is not native: except in two areas it did not succeed in crossing the mountain divide to recolonize the area after the final retreat of the ice. The local eastern race of Norway spruce, and still more the German "race" imported

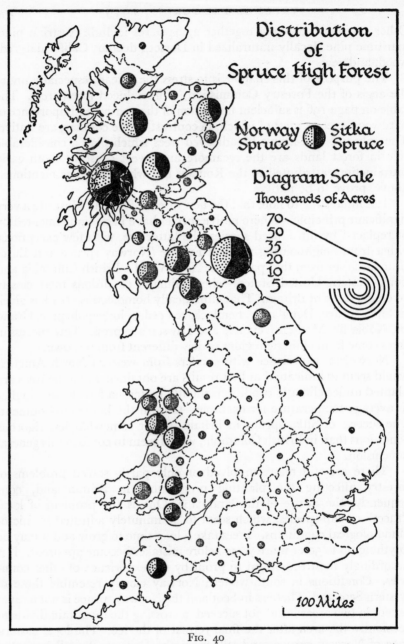

Distribution
of
Spruce High Forest

Norway Spruce   Sitka Spruce

Diagram Scale
Thousands of Acres

70
50
35
20
10
5

100 Miles

FIG. 40
The Distribution of Spruce High Forest in Britain (Census of 1947-49)
The dominance of spruce in the wetter west is clearly shown, and especially of
the Sitka spruce

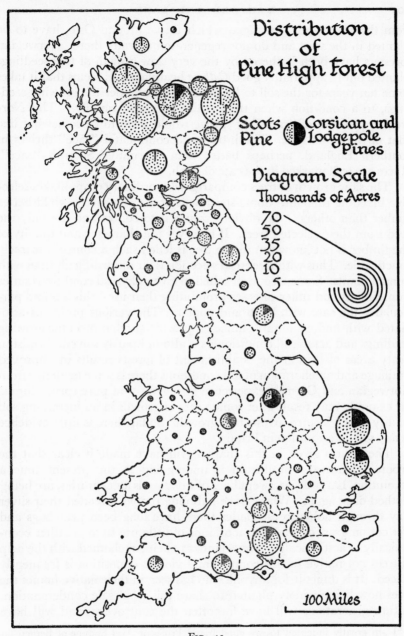

Distribution
of
Pine High Forest

Scots Pine · Corsican and Lodgepole Pines

Diagram Scale
Thousands of Acres

70
50
35
20
10
5

100 Miles

Fig. 41

The Distribution of Pine High Forest in Britain (Census of 1947-49)
The preference of pine for the drier east is clearly shown, and also that the
native Scots Pine holds its own

from the Harz and Thuringia, which thrive around Oslo, have to be planted in the west and do not regenerate naturally there. Norwegian foresters have been bothered by the very slow growth of tree seedlings planted out in apparently good *Calluna* heathland. It seems that it takes some ten years for the soil to be brought back under the growing seedlings, to a condition when trees really go ahead rapidly.* This Norwegian problem of "races" of spruce would seem to be somewhat like that of the Scots Pine. The introduced continental "race" thrives in southern England, perhaps better than the native Scottish "race". Races are well marked in the case of larch.

The slow growth of trees compared with ordinary crops makes selective breeding difficult. Some specimens of a given species yield better timber than others : the obvious course is to collect and use only the seed from these selected trees. If however the character and quality of the timber is not known until the tree is felled such a course is scarcely practicable. Thus with trees seed selection is far more difficult than with annual plants. Some trees well suited to given physical conditions suffer from fungus and other diseases prohibiting their use—this is found particularly the case with larch and poplars. The various problems associated with finding suitable nurse-crops which both protect the growing seedlings and act as soil conditioners is also but partly solved. Whilst in peaty areas of Britain the establishment of forests results in improved drainage and amelioration of soil conditions there is some tendency from Norwegian and German experience to suggest that pure spruce forests for example may result after some years of growth in an increasing soil acidity which in turn arrests tree growth. So far there is little evidence of any soil deterioration in Britain.

The figures given in this chapter will have made it clear that the arboreal landscape of Britain is undergoing at the present time a revolution. Large areas, swept bare of woodlands for centuries, are being clothed once again. Where once babbling brooks threaded their silver way through noble forest glades there have long been peat bogs and wet cotton grass or *Molinia* moorlands of little use to man either economically or aesthetically. These lands are being reclaimed with the help of trees not native to this country and violent opposition is frequently voiced. It is difficult for anyone who has seen in their native homes the trees now being widely planted to share this wholesale condemnation. Future generations will have forgotten the controversy and will have

---

*I am greatly indebted to my good friend Professor Axel Sømme of Bergen for showing me these problems on the ground.

John Markham

Plate 19. SITKA SPRUCE ON THE EDGE OF A FORESTRY COMMISSION PLANTATION
Now the commonest forest tree in Britain this native of western North America has
stiff foliage with a lovely bluish sheen: this view shows the straight edge of a typical
plantation (to which many object on aesthetic grounds) and demarcated by rabbit-
proof fencing   (*pp. 180 and 196-200*)

Plate 20a. SNOWDROPS; *b.* DAFFODILS. Two garden escapes which have established themselves locally.

(Photographs by Eric Hosking)

come to love the new beauty created by this generation, just as we have come to love those strange misplaced clumps of beeches on our bare chalk heights.

Two further tables may be included as of general interest. One breaks down the woodland into types and shows that in 1962 high forest is now two-thirds of the whole, against only half at the Census of 1947.

ESTIMATED WOODLAND AREA CLASSIFIED BY TYPES AND COUNTRIES, 1962
All Woodlands of 1 acre and above, to the nearest thousand acres.

	England		Scotland		Wales		G.B.	
	Acres	%	*Acres*	%	*Acres*	%	*Acres*	%
High Forest								
Mainly Coniferous	687	33	970	59	281	60	1,938	46
Mainly Broadleaved	676	32	118	7	97	21	891	21
Total, High Forest	1,363	65	1,088	66	378	81	2,829	67
Coppice and Coppice with Standards	305	14	—		16	3	321	8
Unproductive	434	21	549	34	72	16	1,055	25
Total Woodland	2,102	100	1,637	100	466	100	4,205	100

The second table shows the share of the State in the total forest.

WOODLAND AREAS CLASSIFIED BY TYPE AND OWNERSHIP—GREAT BRITAIN, 1962
All Woodlands: including plantations

ESTIMATED WOODLAND AREA CLASSIFIED BY TYPE AND OWNERSHIP—GREAT BRITAIN,
1962
All Woodlands of 1 acre and above, to the nearest thousand acres.

	Private Woods		State Forests	
	Acres	%	Acres	%
High Forest				
Mainly Coniferous	677	35	1,261	65
Mainly Broadleaved	753	86	138	14
Total, High Forest	1,430	50	1,399	50
Coppice and Coppice with Standards	293	91	28	9
Unproductive	978	93	77	7
Total Woodland	2,701	64	1,504	36

Although at the 1946-47 Census the State owned only 18 per cent. of the recorded woodland area, it had 30 per cent. or nearly a third of the high forest and over a half (52 per cent.) of coniferous high forest. By 1962 the State owned 36 per cent. of the woodland area, nearly 50 per cent. of the high forest and 65 per cent. of coniferous high forest.

CHAPTER 15

GUESTS — INVITED AND UNINVITED*

THE EVIDENCE from seeds, leaves and pollen in post-glacial deposits makes it possible to say with considerable certainty which trees have been in Britain since prehistoric times. Similar evidence establishes the right of many other plants to be described as truly British but with a very high proportion of our flora the evidence is circumstantial rather than direct. Wheat was certainly cultivated in Britain in pre-Roman times and it is difficult to believe that wheat brought for seed from the Continent was free from the seeds of cornfield weeds. Indeed purity of seeds mixtures has only been achieved with difficulty during the past century. Some samples of imported grass seeds in the last century for example are known to have included up to 68 per cent. of alien weeds. Sir Edward Salisbury† records that of 11,000 samples of Italian Rye Grass examined by Johnson and Hensman prior to 1912 at Seed Testing Stations the average of impurities was 10 per cent. Undoubtedly unintentional introductions due to the movements and work of man have been going on since agriculture began and perhaps even earlier. Thus the delightful blue Cornflower, *Centaurea cyanus,* is probably of oriental origin from the eastern Mediterranean but it had already reached Britain in Neolithic times. Similarly the southern poppy, *Papaver argemone* has been found in deposits of Roman times at Silchester but is not strictly a "native".

Some introductions are known to have been made in the 17th century whilst in the late 18th and early 19th century American species begin to appear. Very rarely is there direct evidence of the date or means of introduction but over the last two centuries botanists have recorded first appearances and the use of first records is at least a rough

---

*I am greatly indebted to Sir Edward Salisbury for much help with the first part of this chapter and to James Fisher in the latter parts.

†A changing flora as shown in the study of weeds of arable land and waste places *Weeds and Aliens,* by Sir Edward Salisbury, New Naturalist, 1961.

guide to the date of introduction. Some aliens which have become firmly established such as the Canadian Fleabane (*Erigeron canadensis*) first recorded in 1690 were early arrivals and must have been accidently introduced since they are not sufficiently attractive in themselves as to have been deliberately brought in for the embellishment of gardens. On the whole alien plants arriving in this country find a well established vegetation and, unless aided by man, competition is too severe for a wide or rapid spread. Thus garden escapes which have really established themselves firmly in the countryside are few (but see Plate 20a and 20b).

If we regard the uninvited guests as those which have arrived in the course of the last two or three centuries the number of plants, other than trees, which we consider separately, which have so settled in and multiplied as to affect appreciably the landscape is quite small. Yet the balance of the British flora is constantly changing. The provision of such artificial environments, previously noted, as canals and canal banks, railway cuttings and embankments, roadside verges and indeed hedge-rows as a whole has permitted a wide spread of previously localized species. The colonization of bombed sites during and after the last war affords a fascinating study of the use made by plants of a peculiar environment suddenly widespread and common.

A very interesting example of rise and fall in the British plant world is afforded by contrasting the spread of the Rose-Bay, or Willow Herb (*Chamaenerion (Epilobium) angustifolium*) with the demise of once familiar cornfield weeds. Linnaeus was particularly unimaginative when he gave the specific name (narrow-leaved) to one of the most flamboyant of our native flowering plants. He lived too soon to appreciate the common American name, "fire-weed" for this plant—so called from the way it colonizes burnt-over forest areas almost before the ashes are cool. Its habitat was noted by Johns in his *Flowers of the Field* as "damp woods ; rare except as an escape". Today it is very widespread and often domin-ates acres of land : it was the most frequent of all colonizers of bombed sites and its spectacular spread elsewhere has been associated with the advent of the motor car—more picnics and more heath fires. By way of contrast the two plants known in Elizabethan days as Corn Cockle (*Agrostemma githago*) and Darnel (*Lolium temulentum*) have with modern cleaning of wheat and other corn become decidedly rare.

Amongst weeds which have spread so as to affect the appearance of

---

Plate XXVII. EDIBLE DORMOUSE
Edible Dormouse (*Glis glis*), a recent introduction, raiding a store of apples in a garden shed, Bedfordshire. (*John Markham*) (See page 216)

Plate XXVIIIa.—CHINA CLAY WORKS, WHITEMOOR, CORNWALL
Vast mounds of glistening white quartz waste dot the wide moorlands over the granite
masses of Devon and Cornwall

b.—GLASSHOUSES IN THE LEA VALLEY, MIDDLESEX. Although a form of intensive agriculture, cultivation under glass leads to great factory-like construction

the countryside of Britain in a marked degree may be noted the Pineapple weed (*Matricaria matricarioides*), a native of Oregon first recorded in 1871 but only found rarely in widely separated localities until after 1900. The seeds are distributed in mud : the favoured habitat is waysides and road verges. The perfect dispersal agent is the motor car : mud clinging to tires has made this one of our most widely distributed and commonest weeds. The larger the tire and the more mud the better: hence the efficiency of the motor lorry in the First World War.

Plants seem to share with human beings preferences for contrasted modes of transport. The Oxford Ragwort (*Senecio squalidus*) remained long faithful to the railways, preferably the Great Western, though it has tended to desert the railways since their nationalization. A native of Sicily it was first recorded at Oxford in 1794 probably as an escape from the Botanic Garden. It did not however arrive in London until 1867 having apparently travelled along the waste ground by the side of the railway. It was slow to desert railway sites until its great spread in the Second World War over more than half all blitzed sites. The Sticky Groundsel (*Senecio viscosus*) is a near relative of the Oxford Ragwort : it too has spread widely since its first appearance (recorded 1660) from southern Europe. It seeks waste ground and the increased speed of gravel working has afforded many new sites—sites which are favoured also by the two species of wild lettuce, *Lactuca virosa* (1570) and *L. scariola* (1632). Both these were still rare a century ago. The Canadian Fleabane also grows on similar waste land and was found by Salisbury on 40 per cent. of London's blitzed sites. Thanet Cress (*Cardaria draba*) which came from south-eastern Europe via the Island of Walcheren into the Isle of Thanet in 1809 and the Beaked Hawk's-beard (*Crepis taraxacifolia*) which also infests waste land are other examples of now widespread aliens.

Other alien weeds remain more local. *Claytonia perfoliata* is conspicuous in Breckland and is an American species first recorded here in 1852—rather later than its colleague *C. alsinoides* (1838). An American species of *Oxalis* with tiny bulbs has become so pernicious a weed on light land at Bude (Cornwall) as to cause allotment holders to give up their plots.

Sir Edward Salisbury has called attention to the very special interest of "Gallant Soldiers" (*Galinsoga parviflora*) which is a native of Peru and escaped from the Royal Botanic Gardens at Kew. Its seeds are dispersed by wind but the pappus is not very effective and thus provided evidence of the importance of upward convection currents for wind

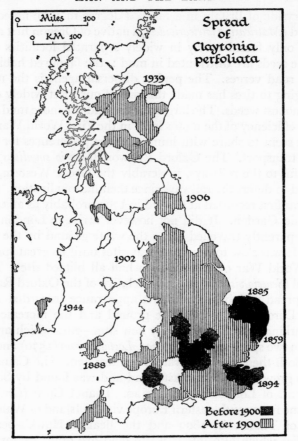

FIG. 42

Map illustrating the spread of an alien plant (*after E. J. Salisbury*)
Claytonia perfoliata was first recorded in 1852 but the spread only became
rapid after 1900

dispersal as its rate of spread over the bombed sites of the London area
was shown by Salisbury to correspond almost exactly with what is now
known to result from heavy bombing.

The plants already mentioned may be classed as weeds. A few conspicuous wild flowers are aliens and two at least are noteworthy. The
Yellow Monkey Flower (*Mimulus guttatus*) hails from America and has
become widespread by British streams and ponds. The Indian Balsam
(*Impatiens glandulifera*) now grows locally as great masses six to eight feet

Plate 21a. A Rabbit Burrowing under a Hedge

       b. A Grey Squirrel

Examples of two animals not native to Britain which have established themselves very firmly—the rabbit for many centuries, the grey squirrel in the last half century *(pp. 211-13 and 210)*
                                                             (Photographs by John Markham)

Plate 22a. A BLACK RAT
An alien which probably arrived with returning Crusaders (*p. 211*)

b. MAN SHOOTING COYPU RATS IN EAST ANGLIA
All the rats have been introduced into Britain (*p. 215*)   (*Photographs by John Markham*)

high by such streams as the Avon at Bath or the Tweed in Berwickshire.

Turning to shrubs a great contribution to the British landscape has been made by *Rhododendron ponticum* now widespread on sandy soils. It is a native of the Caucasus and the Balkans and referring to its spread in the Killarney woods of south-western Ireland, Tansley calls attention to the rare phenomenon of the successful invasion of an undisturbed community by an alien.

Whilst the Rhododendron has boldly attacked and at least partly conquered native strongholds another flowering shrub has made use of openings accidentally provided by man. *Buddleia davidii* has become familiar on many blitzed sites and waste ground, especially round London.

A rather special example is that of *Elodea canadensis*, the Canadian Water weed, which threatened to choke many of our canals and sluggish streams and to exclude many native species. In some years, despite expensive cleaning in spring, it has formed such a solid mass in the canal at Bude at to prevent summer boating by visitors : in other years it scarcely survives. It may be too that the period of great abundance which characterised its first appearance has been followed by one of general decline in the country as a whole.

A truly remarkable story is that of the salt-marsh rice-grass, *Spartina*. There is a native British species *S. stricta* which plays a subordinate part in some of our coastal marshes and another species *S. alterniflora*, a native of North America. A natural hybrid between the two is *S. townsendii*, a tall, strongly-growing perennial grass first noticed in England along Southampton Water in 1870. It has since colonized deep mobile mud, washed by every tide, especially in parts of Southampton Water, in Poole Harbour and elsewhere along the southern coasts and has also been planted elsewhere. As Tansley observes, "No other species of salt-marsh plant, in north-western Europe at least, has anything like so rapid and so great an influence in gaining land from the sea." It forms almost pure "meadows", binding the mud already accumulated by its roots and catching and holding more drifting in by every tide. A rough estimate based on Ordnance Survey maps shows that half a million acres round the coasts of England and Wales lie between high tide and low tide marks : eliminating sand and shingle much of this is mud and doubtless a large proportion could be colonized by *Spartina* and later reclaimed.

Over all those parts of Britain where spontaneous or sub-spontaneous vegetation is dominant—the mountains and moorlands, waysides and

ML.—P

wastelands, stream courses and cliffs—there have been and still are great changes in the floral composition of the plant assemblages. On the whole, however, it is a changing emphasis amongst native plants, such as the spread of bracken or rose-bay, rather than successful invasion by aliens.

In the animal world, by way of contrast, the alien has been more successful. The elimination of the large wild animals and birds of prey provided an environment where natural enemies of certain visitors were almost non-existent. In this rabbit paradise, the human rabbit trapper was often almost the only enemy, with the result that natural checks following some rabbit-Malthusian principles not yet understood came into existence. Britain has proved a pleasant home for the grey squirrel, *Sciurus carolinensis*, an alien from eastern North America, which escaped from zoos at least as early as 1830 or was deliberately liberated in several places (notably Woburn) between 1889 and 1929. Its attractive fur coat hides a villain of the worst order. It eats almost anything, with a preference for nuts, seeds, young main shoots of valuable trees, birds' eggs, young birds—and adults if it can catch them. In the brief space of thirty years squirrels from the Woburn centre alone spread over 1350 square miles. In recent years there has been such a rapid increase that 50,000 a year are destroyed by various means in State forests alone. If this handsome fellow retains public sympathy let it be broadcast that he is steadily replacing the lovely little native red squirrel—though exactly how is not yet clear (Plate 21b.).

The primitive home of cave-man must have had many disadvantages but one modern plague at least was absent. Apart from an occasional visit from a field mouse, rats and mice were unknown. They are a disease of civilization and have adapted themselves almost unceasingly to the habitats and food provided by man.

The whole story of the house mouse (*Mus musculus*) is a fascinating one—not least curious being the parallel to be observed between mice and men. The town mouse cannot be induced to move far from the shelter of the urban home. The country mouse on the other hand prefers the greater freedom of farm and barn. The more venturesome have taken anew to life in the fields. Each group has developed distinctive traits and to untangle the complex race-relationship has defeated the ablest biologists.

Mice are destructive enough, but rats are enemies of man in many more ways—in addition to destroying quantities of food, they contaminate much more and in addition spread disease. The dreaded plague, of course, is spread by rat-fleas. Like the house-mouse the two species of rat—the brown and the black—rarely move far from habitats kindly provided by man. The rat which proverbially leaves a sinking ship is the black rat (*Rattus rattus*) : its distribution ashore is largely restricted to ports and it rarely deserts the shelter of buildings. The fleas of the black rat carry the murine form of plague. The animal seems to have arrived in Britain as a stowaway sometime in the 12th century, probably in the ships of the returning Crusaders (Plate 22a.).

The brown rat (*Rattus norvegicus*) has become ubiquitous on land but rarely goes to sea. Surprising as it may seem this native of the Caspian did not begin its large scale invasion of western Europe until 1727. It reached Britain, probably in ships trading with Russia, in 1728 or 1729 and so quickly made itself at home that within forty years the trade of the ratcatcher was a thriving one. It has spread all over the country : only in the ports is the black rat still dominant though it took the brown rat 150 years to oust the black from country districts. During the Second World War 80 per cent. of rats trapped in the Port of London were black rats and this clever climber still haunts the West End. Almost everywhere else brown rats predominate. As with the house mice, there are town races and country races : in the country rats will often live in the fields and hedgerows in summer seeking the shelter of haystacks and corn-ricks or barns in winter. There is only one surviving colony of country black rats—on Lundy. The seriousness of food losses due to rats during the Second World War led to intensive research into their numbers, habits and the best methods of combating them.

Of all the guests, invited and uninvited, none made himself more at home in Britain than the rabbit (*Lepus (Oryctolagus) cuniculus*) which is believed to be originally a native of the western Mediterranean basin. Although the rabbit has many enemies especially amongst the smaller carnivores and prey birds, those enemies are the very creatures which civilization and the spread of man's activities tended to eliminate, making the world a much more pleasant place for the busy rabbit. The female rabbit is old enough to bear young at six months, *can* produce litters at monthly intervals and *does* produce several a year, each litter comprising three to eight (usually five) young. It can live for seven or eight years, though the expectation of life is much shorter. Thus a single doe *may* produce over a hundred young. An average gestation is 28 to 30 days, but the female rabbit normally mates again within 12 hours of the birth of a litter.

It is not clear when rabbits were first introduced into Britain—perhaps by the Normans*—but they became the most frequently seen of all wild animals. Their spread by ill-timed introduction into Australia and New Zealand is so well known as to be notorious: it is not perhaps as commonly recognized that until 1820 the rabbit (or coney, to use the old name) was almost unknown over much of Scotland. Though mentioned from Aberdeen in 1424, the great spread was not till the end of the 18th century. Similarly introduction into Ireland according to Moffat was in or before 1282 but the spread was quite late. Although no census was ever attempted of the wild rabbit population of Britain, estimates can be based on the numbers trapped and consigned by railway to urban markets. In the year 1948 three counties alone in western Wales consigned 4,500,000 rabbits in this way; Devon and Cornwall over 1,000,000. Harrison Matthews accordingly hazards the suggestion that "the rabbit population of the country must equal or exceed the human one in size". No wonder the Ministry of Agriculture considered 50,000,000 rabbits far too many, but the farmer on the marginal lands was apt to regard the cheque in his hands from rabbit sales as more tangible than the cheque he might receive from crops he might grow if rabbits were eliminated.

Unlike the hare, the rabbit is born naked and blind in a fur-lined burrow which gives protection against the vagaries of climate. Rabbits have long been enjoyed as food in the country and it was the fashion to establish breeding "coneygarths" or "rabbit-warrens" in the more sandy parts of large estates. William Harrison in his description of Elizabethan England first published in 1577 devoted a chapter to "parks and warrens" and refers to the post of "warrener". "As for warrens of conies", he says, "I judge them almost innumerable, and daily like to increase." He refers to the trade in young rabbits for the London market and of buying of "older conies" from further off for breeding. He states that black skins were regarded as far more valuable than grey and "this is the only cause why the grey are less esteemed". It would seem therefore that the rabbits bred in warrens may have been of domestic breeds differing from the wild rabbits later found everywhere in the open though blacks are frequently known among wild rabbits.

The toll on growing crops and corn taken by rabbits became so enormous that extensive operations to destroy such colonies by use of poison gas were frequently undertaken. Though this supported efforts towards greater food production, it was not whole-heartedly endorsed

*In 1176 there were rabbits in the Scilly Isles (see Veale, 1957).

by farmers, especially in the poorer and remoter parts. Many a small farmer found his chief pleasure in life going out on a slack afternoon with a gun under his arm and "potting" at rabbits. The resultant bag served to augment the supplies of the table, and the sale of rabbits for distant urban markets was a useful source of cash. It was difficult to persuade the farmer that the cash return was actually considerably less than the value of the grain or other food the rabbits took from his farm or the damage they had done. Shooting over one's own land has a well-nigh irresistible appeal to many a countryman with the result that a farm worker who could scrape together a little capital to buy or rent his own small holding readily gave up a better financial and more secure position at a regular wage for this reason alone.

Whilst the rabbit in Britain never became the menace it was in Australia, New Zealand and elsewhere, it was undoubtedly one of the most persistent and destructive pests of the countryside. Rabbit-proof fencing must not only be high and close, but sunk deeply in the soil because of the burrowing habits of the animal. Yet no plantation of forest trees could be made until a rabbit proof enclosure was provided, often at an alarmingly high cost (see Plate 21a).

In 1952-3 the fatal rabbit disease, myxomatosis, swept across France and practically eliminated the rabbit over large areas. It reached Britain apparently in the late summer of 1953 and was recorded first at Edenbridge in Kent, shortly after at Robertsbridge in Sussex and near Lewes. The disease is caused by a virus which originated as far as known in domesticated rabbits in South America and was kept alive in laboratories in Europe. It was actually used deliberately in clearing areas of rabbits with much success in Australia. It was so used by the owner of a woodland estate in Sweden but public opinion—outraged by the suffering caused by the disease—compelled him to desist. It causes intense swelling of the face and hind parts, exudation of a highly contagious fluid, and blindness. The animals come out to die in the open and the sight is far from pleasant; moreover the disease is spread by wheels of cars. It is probably in this way that it was introduced into England. Though fortunately the disease seems restricted to rabbits and does not even affect hares, the domesticated rabbit is especially susceptible. Whether the disease is to be regarded as a blessing or a curse was hotly debated in France. In that country both the *chasseurs* and the peasant farmers were deprived of their sport and a popular food supply and formed associations for the defence of *les lapins sauvages*. The first action of the Ministry of Agriculture in this country was to set up a

Myxomatosis Advisory Committee and to attempt to control the disease. But the disease ran its course and the rabbit was almost ex-terminated throughout the country. A few that survived adopted at least temporarily the habit of nesting above ground. Official policy is to prevent re-establishment.

Reference was made above to 'rabbit malthusianism'. It only latterly became known that there was a form of natural birth control in that a large proportion of the rabbits conceived were never born but died within the womb and were reabsorbed. Rabbit influenza and rabbit syphilis also killed off large numbers.

In addition to meat, the wild rabbit yielded fur which under the trade name of coney was very widely used, skilfully worked by trimming and in other ways into several types of fur, for the cheaper fur coat.

As with so many other mammals and birds, it is impossible to say when the rabbit was first domesticated. The Chinese certainly had domesticated rabbits in the time of Confucius. The breeds which have been developed in captivity have an enormous range in type: indeed domestic breeds of rabbits show a greater variety than any animals except dogs. Some types have been developed for meat—for example the great Belgian hare—others are purely for show.

Unlike the rabbit, the two species of hare are native to Britain.

The muskrat or, to use its more aristocratic Red Indian name, the musquash (*Ondatra zibethica*) affords the perfect example of the guest who stayed to dinner—with disastrous results. Three-quarters of all the animal skins entering the fur trade of Canada in many years are musk-rat skins and this fact alone illustrates the widespread use of a hard-wearing and attractive fur. Consequently muskrat "farms" were established in many parts of Europe, including Britain. Being natural wanderers and great burrowers it is not surprising that there were escapes from the fur farms starting with 1927 in Scotland and 1929 from Great Missenden and Louth. The animal, a very rapid breeder, became naturalized in the next few years in many parts of England, Wales, Scotland and Ireland. In Canada there are vast marshlands where the muskrats can burrow interminably, lay up winter stores as mounds of reeds and rushes to form lodges into which they later burrow and eat. No great harm is done if undermined trees come crashing into the morass. But in a small country such as Britain such activities were disastrous. Canal and river banks especially of the Severn were undermined so that water escaped and flooded low-lying meadows: streams were blocked by fallen trees and the water spread out into extensive marshes. In Britain's mild climate, with few if any enemies, breeding and feeding can go on most of the year. Each female will produce six or seven litters of eight—say 50 young—each season. Some known escapes took place in Perthshire in 1927; within five years there was no doubt of the serious damage being caused. The keeping of muskrats was controlled by licence in 1932 and prohibited the next year. The Department of Agriculture for Scotland and the Ministry of Agriculture and Fisheries undertook a trapping campaign, especially in the winter of 1933-34. Nearly a thousand were caught in Scotland alone (together with nearly 7,000 innocent and unintended victims of the traps) and many more in England. Similarly in Ireland 487 muskrats were trapped in the winter of 1933-34—the progeny of a single pair which had escaped a few years earlier. Fortunately the campaign was successful: by the end of 1939 the muskrat, after ten years of freedom in Britain, was extinct as a wild guest of our countryside. As Harrison Matthews has well said, "A person releasing a pair of muskrats does far more damage, and causes far more trouble and expense in clearing up the results of his action, than a person releasing a pair of man-eating tigers."

Another fur-bearing animal is the two-foot long South American coypu rat (*Myocastor coypus*)—the nutria of commerce. There were escapes from the fifty fur farms established between 1929 and 1939 and colonies established themselves in East Anglia. As numbers increased they undermined the banks of water courses and caused damage to crops, especially sugar-beet. A government campaign had to be undertaken for their elimination and it is said 100,000 were killed in 1961.

As James Fisher has observed "a new and sinister possibility is the escape of mink from British fur-farms and the establishment of this aggressive and opportunist member of the weasel family as a naturalized wild animal." This has happened in Iceland and, as Fisher observes, such a locality as the Spey valley would suit it well with consequences to trout, char, ducks and wild fowl too horrible to contemplate.

Of recent years we have invited a number of guests from amongst the deer. It is impossible to separate the indigenous and introduced strains of fallow deer and the same is true of the Siberian race and the native race of the roe deer. Two races of the Asiatic Sika deer are now feral, i.e. have become established as wild, also two races of the Asiatic barking deer (Muntjac) and the little Chinese water deer and the Indian chital or axis deer. The larger American blacktailed deer is now at large in at least one area of England. Thus in addition to the three native deer, six or seven are now wild in Britain having escaped or been released, especially from Whipsnade or Woburn. The latest is the reindeer—a Government-sponsored introduction into Scotland. So far there are no complaints, but one wonders.

Another interesting guest is the Fat or Bushy-tailed Dormouse, an animal as large as a squirrel, introduced in 1902, and now at home in the Chilterns. The Romans fattened them as table delicacies : it is possible they may become pests as they are in some continental orchards and vineyards. At present the chief complaint is with regard to their noisy night life for which they invade the attics of homes along the Chiltern scarp. (Plate XXVII.)

BIRDS

It is calculated that more than ten per cent. of British breeding mammals, birds, reptiles and amphibians are aliens—introduced by man but now feral or wild. Amongst birds there are various species of pheasant (the common pheasant is credited to the Romans), the famous reintroduction of the capercaillie, the French partridge, many ducks, the Canada goose, Barbary dove, Little owl and many others. Perhaps the Muscovy duck, peafowl and swans should have been included under Farm Animals in Chapter 11, but they call at least for some mention.

PEA FOWL

Although classical authors allude frequently to the high esteem in which peacocks were held as table birds so that a banquet was scarcely complete without a peacock served up as a main dish surrounded by its brilliant plumage, the bird can no longer be regarded as entering into food supplies. No longer an inhabitant, if it ever was, of the poultry yard it is becoming scarce even as an ornamental occupant of the pleasure grounds of erstwhile great houses. The peacock, *Pavo cristatus*, has never become completely domesticated and retains most of its wild habits : it is not a prolific breeder and its destructive habits in ordinary gardens combine to cause a decrease in numbers. The magnificence of the feathers in the spread tail of the male could scarcely fail to attract attention as it did of King Solomon (1 Kings, X,22, 2 Chron. IX, 21) and of the Greeks and Romans so that it was introduced westwards at an early date from its home in India. Peafowl of an allied species are still common in the forests of Burma and the peacock has been adopted as the symbol of the Republic of Burma.

It is interesting to note that two varieties have developed in the semi-domesticated birds. One is an albino form, more or less pure white ; the other is the so-called "Japanned" Peacock (not Japan Peacock) with brilliant lustrous blue feathers over the upper wing covers.

MUTE SWAN (*Cygnus olor*).

It is a commonplace of the literature of domestication that the mute swan, Britain's largest flying bird, is an introduced and semi-domestic species. However the researches of N. F. Ticehurst (summarised in Witherby's *Handbook of British Birds*, vol. 3 : 177-78) show that the bird was almost certainly indigenous in eastern England and probably on the Thames, and that it was simply brought into semi-domestication, by pinioning, marking and redistribution, shortly after the Norman Conquest, and gradually ceased to exist as a wild species. Ticehurst shows that from the thirteenth to the eighteenth centuries swans were " Birds Royal," the property of the Crown, and that some populations were leased by the Crown to corporations and landowners, who marked their birds in distinctive ways.

In England swans were formerly far more numerous than at present and the young or cygnets (actually a table name derived from Norman French contrasting with swan, which is Anglo-Saxon) were highly esteemed for the table. By the reign of Elizabeth I the privilege of ownership had been widely extended and some 900 separate "swan-marks" were recognized by the Royal Swanherd, whose jurisdiction extended over the whole country. At the present day there are still Royal Swans on the Thames and in August is the ceremony of "Swan upping"—the taking up of young swans and marking them. The English swan-laws and regulations, together with the complicated system of swan-marks, form a fascinating piece of the country's legal and social history.

Since the eighteenth century the Crown has largely ceased to enforce its traditional rights, and the marking of swans is preserved for the sake of tradition only on the Thames and in Dorset and Norfolk : elsewhere the populations have reverted to a truly feral state.

For many decades there has existed one great Swannery in England —that on the sheltered waters known as the Fleet behind Chesil Beach in Dorset. In the early eighties Lord Ilchester is said to have owned there between 700 and 1,400 birds ; in recent years the population has fluctuated between 500 and 1,000.

Elsewhere swans usually establish themselves wherever a quiet river, canal or lake with appropriate shelter suggests a home. A pair of swans will take up possession of a stretch of water, build an ungainly nest of aquatic plants and sticks, perhaps a couple of feet high and three to six feet across, on the top of which in a slight hollow some four to nine greyish-olive eggs are laid. Some swans are great exhibitionists and deliberately build their nest in a conspicuous and frequented spot later displaying their cygnets with great pride (Plate 23a). On the River Cam for example swans frequently nest near one of the Cambridge bridges—a pair did this in 1942 a few feet from Silver Street Bridge, observed by thousands of passers-by daily. In 1941 a pair built their nest by the side of the old canal in Bude a few yards from the sentry's box on a military vehicle park. Lorry drivers had to be careful not to back on to the nest and a disgusted Cockney soldier on sentry duty was heard to exclaim, "Nursemaid to a blinking swan, that's me."

THE INVISIBLE HOST

Of necessity this brief account of guests from the animal world must be restricted to those larger creatures adorning or despoiling our countryside. But there have been many smaller creatures, amongst the uninvited guests, including for example literally hundreds of insects. The common cockroach (*Blatta orientalis*) a native of tropical Asia arrived some time before it was first recorded in 1634 and gradually spread from the ports. It has been followed in more recent times by relatives from Germany (*Blatella germanica*), America (*Periplaneta americana*) and Australia (*Periplaneta australasiae*) still of restricted distribution. The near East sent the bed-bug to Britain in the 16th century, but more difficult of control have been the plagues of the countryside such as the woolly aphis (*Eriosoma lanigera*) causing American blight on fruit trees and first recorded in 1787. The dreaded Colorado beetle, so destructive of potatoes, reached France about 1922 and every effort was made to keep it out of England. A colony near the lower Thames at Gravesend-Tilbury was apparently eliminated in 1933, but the beetle later reappeared. When the Ministry of Agriculture offered rewards for any specimens caught in Britain, it overlooked the incentive to matchbox smuggling from the Continent thereby induced.

Just as introductions of the larger animals raise many and often unexpected problems, so certainly does the introduction of apparently harmless creatures of the lower orders.

CHAPTER 16

THE NINETEENTH CENTURY

Aᴸᵀʜᴼᵁᴳʜ the changes in land use were briefly considered in Chapter 7, it is necessary to return now and consider them in slightly more detail. At the present time we are concerned as a nation with the more efficient use of our land resources. Planning is always dangerous unless it is founded on a firm factual knowledge both of the present and the past, especially the immediate past.

It is a sad commentary on the life of nations that nothing stimulates change and development so much as war or the threat of war. Towards the end of the 18th century, when Napoleon Buonaparte was making himself master of Europe, the threat of invasion became a real one to Britain. Indeed, Napoleon's elaborate plans for the invasion of the islands were only finally abandoned after his naval defeat by Nelson at the battle of Trafalgar in 1805. The invasion threat had led to the construction of those remarkable round towers, many of which still stand around the south-east coast, the Martello towers. Of more lasting note was the institution of the Ordnance Survey for the purpose of preparing maps for the use of the army. To this day our national surveying and map-producing organisation, though it comes, oddly enough, under the Minister of Agriculture and Fisheries and is concerned primarily with the supply of maps for civilian use, is still staffed at its higher levels by serving officers of the Royal Engineers, and retains the name Ordnance Survey. British maps are still to a considerable degree influenced by military requirements : they are closely contoured and pay attention to careful delineation and colouring of roads, and presence and type of woodland. Unlike maps, for example, of the Netherlands no attempt is made to show character or use of farm land on maps of any of the standard scales. The existence of open, unfenced moorland or rough pasture is obviously of military importance : it was clearly shown on the first maps issued from 1801 onwards on the scale of one inch to one mile. This was the standard scale of the first maps issued by the Survey—the beautiful and accurate steel-engraved First Edition, showing relief by elaborate hachuring. In the last half of the 18th century many excellent county maps had been issued by private firms. The beautiful maps of John and Ann Rocque notably for Shropshire (1752), Middlesex (1754) and Surrey (1762) led the way and were followed by those of William Faden for Norfolk (1790-94) and London (1802), Andrew Armstrong (notably Lincolnshire, 1779) and C. and J. Greenwood (notably Yorkshire, 1818 and Shropshire 1827).

Many of these maps showed some details of land use such as woodland, moorland and marshes. Though some were also issued, especially the Greenwood maps, after the publication of Ordnance Survey sheets, which they seem to ignore, they lack the accuracy of the official maps and were destined to be superseded by them.

Thus, two important features of land use, woodland and moorland, are shown accurately on maps dating from the earliest years of the 19th century. This has an added importance because it was just at this time that the Board of Agriculture and Internal Improvement was issuing its county reports. As already noted this Board, though it may be regarded as a natural parent of the Board of Agriculture instituted as a government department much later, was an independent body supported by public funds. Its first chairman was Sir John Sinclair, its energetic secretary the agricultural writer Arthur Young. His rival as an agricultural writer William Marshall had, it is said, hoped to be appointed secretary. When he was not so appointed he devoted much energy to sustained criticisms. One of the first acts of the Board, on its establishment in 1793, was to arrange for reports, county by county, on the state of agriculture. These were issued in quarto volumes with very wide margins and distributed so that as many comments as possible might be collected. Then followed the publication of the definitive work as an octavo volume. Sometimes it was the earlier report revised, more often the earlier enlarged or even entirely re-written, and in some cases by a different author. Each volume is entitled *A General View of the Agriculture of* . . . . and most add hopefully, "with observations on the means of its improvement." The publication of these county volumes went on for some twenty years : Scotland was also included. Naturally the reports are unequal in quality but they form together an almost inexhaustible mine of information. Each includes, generally as a folding plate, a map of the soils of the county, and in many instances this remains the only soil map extant. In due course Marshall published a series of volumes analysing the reports with the declared object of rescuing from oblivion those parts which he regarded as worth while. Most researchers prefer to go to the originals.

If we take together the Ordnance Survey maps, the earlier county maps where such are available and the Board of Agriculture Reports*, we are forced to the view that in very many respects remarkably little change has taken place in the past century and a half in so far as the general land use pattern of rural Britain is concerned. The moorland and rough grazings of today occupy very much the same areas as they did at the opening of the 19th century. Since their distribution and extent are determined in the main by unalterable physical factors of elevation, slope, soil, drainage and climate this is not, or should not be, surprising, but the strength of the physical factors in thus limiting land use is not often fully appreciated. As the years have passed by, lowland mosses and marshes have been steadily drained and reclaimed, at times the mountain moorland edge has been pushed back, only to be re-established at its former position at other times. On the whole, forest and woodland remains where it was in its attenuated fragments. The network of roads connecting the long-established hamlets, villages and county towns has changed but little in pattern. Except in those areas where enclosure had not yet taken place, the modern field pattern (though not shown on the First Edition Ordnance Survey) had from other evidence already been established : modern farm units often scarcely differ from those already established.

Of course there has been a complete change in industrial districts where the urban-industrial expansion has obliterated the old rural pattern : elsewhere the rural pattern was certainly established in its main features and in much of its detail.

The wealth of material which exists for the reconstruction of the changing history of land use through the 19th century has never been fully used. Of course the general picture of change is well known. Intensity of home food production with prices attractive to the farmers suffered a severe blow after Waterloo, in 1815. It was shortly afterwards, from 1821 to 1832, that William Cobbett undertook his famous journeys through the length and breadth of the land, and in his *Rural Rides* has given us one of the classics of the English language. Beauty is in the eye of the beholder and Cobbett found beauty in a well-tilled landscape, in fat well-woolled sheep and contented cows. To him there

* Crop returns for 1801 were also collected by parish priests. See H. C. K. Henderson (1952).

was nothing to be admired in a stretch of common or a sweep of hill moorland. The landscapes he described have for the most part a familiar sound today : the most productive cornlands are those which today are classed as No. 1 and No. 2 lands in the National Scheme of Land Classification (see page 244). There have been some major changes : much of Fenland which today is the most continuous stretch of ploughland in England was then fine grazing with a great sheep population.

Fortunately it is possible in many parts of the country to trace more exactly changes in land use. From 1845 onwards the Tithe Commutation Commission was engaged in making an exact survey of all lands still subject to tithe. The Commission had to make its own maps on a large scale, since the Ordnance Survey had not yet started on its large scale surveying. The maps were in manuscript, one copy was deposited in the parish with the parish priest, another at Diocesan headquarters in care of the Bishop and a third at the Commission's central office in London. The first copy is frequently missing and the second may be difficult to trace, but at least one has survived everywhere. Not only is every field delineated but the schedules give the owner or occupier of each and its use at the time. One can therefore reconstruct a land use map and frequently a farm boundary map of the period—rather over a century ago. The work is slow but rewarding: it has been carried out for large areas now in many counties, including Sussex, Surrey, Middlesex, Essex, Bedfordshire, Hertfordshire, Derbyshire and Devonshire. The proportion of arable was high but once again the stability of the land use pattern is demonstrated.

Public concern at the state of farming led *The Times* to send on an extensive tour James (afterwards Sir James) Caird in 1850-1 and he wrote a careful account as a result. About the same time the Royal Agricultural Society instituted a series of prizes for essays on the farming of selected counties. The first was published in Volume V of the Society's Transactions (1845). In the course of a number of years essays were published covering all the counties of England and Wales. For Scotland different information is available. In the 18th century Sir John Sinclair had organized the old Statistical Account— every parson being induced to write an account of his parish. This record relates to the period 1790-1799. From 1834 onwards a New Statistical Account was undertaken, the published volumes being dated 1846-49. In Scotland it was the Highland and Agricultural Society (which had already done much valuable work since its foundation in 1783) which encouraged the preparation of prize essays on the counties.

The eighteen-forties witnessed the failure of the potato harvest in Ireland, the terrible famines which resulted and the initiation of large-scale migration to America. The Government of the day was induced to collect annual agricultural statistics for Ireland (beginning 1847) and, after a lapse of nearly 20 years, a comparable system was initiated for England, Wales and Scotland. The year 1866 saw the farmers of Britain filling up for the first time the annual returns of crop acreage, numbers of livestock, which they have been required to do ever since. The details normally refer to the position on 4th June of each year, hence the familiar reference to the Fourth of June returns. For the first few years the returns were not fully complete but since 1870 the record in this country may be regarded as both complete and having a high degree of accuracy. The returns are made to what is now the Ministry of Agriculture, Fisheries and Food and are confidential. Provided there are more than one or two farms in a parish, the parish figures may be disclosed to the public but are not published. Since many farms may overlap the bounds of a parish there are discrepancies and so the statistics do not form a substitute for a detailed survey and map record.

It was about 1870 that the Ordnance Survey started on its great work of detailed survey and publication of maps on the scale of 1 : 10,560, or six inches to one mile. These are maps on which every field is shown and for some years (until discontinued for reasons of "economy") the surveyors' field notebooks included a column in which to record the use of each field or parcel of land. Where these records exist it is possible to construct a land use map for a very interesting period. It was in the early seventies that British farming was at its peak. It was a period of high farming with large home markets in the ever-expanding industrial towns. True the Corn Laws had been repealed in 1846 and grain from overseas came in duty free, but it was not until the later seventies that the trickle became a torrent, sweeping cheap food into the country in such quantities that it threatened to engulf and extinguish home production.

The official figures show the consequences—the steady drop in arable land and all crops, the rise in the acreage of permanent grass, the abandonment of marginal lands and the consequent rise of rough grazing acreages.

Although the effects of the inflow of cheap food in the latter part of the Victorian era have often been discussed in so far as Lowland

Britain is concerned, the changes occasioned in Highland Britain—Wales and Scotland—have received less attention. My friend and colleague Peter Scott has pointed out the real meaning of the changes in the stocking of Welsh mountain farms studied by Mr. R. Owen of Cresor*. A good example of the changes is afforded by the farm of Owastadannes which extends from the bottom of Nant Gwynant to the top of Snowdon :—

A.D. 1569	A.D. 1789	A.D. 1946
48 cows	15 cows	2,000 sheep
25 heifers		
16 calves	13 calves	
2 bulls	1 bull	
56 sheep	357 sheep	
22 lambs		
19 work oxen	10 steers	

In the early days such a farm afforded the perfect example of a balanced peasant economy. With the Industrial Revolution came the emphasis first on wool and store cattle but, as competition from overseas supplies of wool and meat developed, the farm became just a great hill-sheep run. This story is repeated over much of Highland Britain (with the consequent changes in the floristic composition of hill grazings studied by Wylie Fenton) and many believe that the time is ripe for a return to a more balanced type of farming.

These changes in land use did not take place without effect on the rural population. Whilst, for reasons discussed below, the numbers of farmers remained about constant, the numbers of farm workers declined steadily. In England and Wales male workers over 21 years of age dropped from 676,000 in 1871 to 383,000 in 1931. Though there remained and still remain many rural slums, the farm workers who stayed in the country were better housed and the terrible conditions of many country cottages, of which Cobbett had complained so bitterly, were ameliorated. Where the swing over from arable to grass was very marked, a few shepherds or stockmen took the place of a much larger number of ploughmen. But nowhere in England do we find the conditions which caused such changes and gave rise to so much bitterness as in the Scottish Highlands. There, in the narrow glens, the people eking out a bare existence as crofters, cultivating a patch of ground and pasturing a few poor animals on the hillsides, were

*Mentioned in Scott, R. (1948) *Snowdonia*, p.372 et seq.

Plate 23a. SWANS NESTING, RIVER LEA. Illustrating the very public situations in which swans nest (*p. 218*)

b. THE HEAD OF A TURKEY
An introduction from North America (*p. 155*)   *(Photographs by Eric Hosking)*

Plate 24a. OPEN CAST IRON ORE WORKING
This is an example of good land restoration. The natural surface of the ground is shown on the left, the iron ore is being worked in the centre. On the right the land has been returned and is growing crops.

b. THE BISHOP'S PALACE, WELLS
A good example of man-made beauty which we seek to preserve—the building, wall, moat and introduced horse chestnut.

(Photographs by L. Dudley Stamp)

numerous settlements which had existed from early times. In years of bad harvest there was much distress and even starvation : if food were not distributed by charitable landowners many died of starvation. A specific example of what happened is afforded by the county of Sutherland. There the first Duke of Sutherland, having been advised by a group of agriculturalists that country which might support sheep could never support human beings at an adequate level, determined to move the hardy mountaineers from their inland glens to the sea shores and more accessible valleys. Peaceful persuasion failed and compulsory migration had to be undertaken. Removals started in 1807 and were completed in 1820. And so :—

"The land that once with groups of happy clansmen teemed
Who, with a kindly awe, revered the clan's protecting head
Lies desolate, and stranger lords, by vagrant pleasure led
Track the lone deer, and for the troops of stalwart men
One farmer and one forester people the joyless glen."
                                        (*W. Smith*)

Philanthropy or callous eviction as suggested by this quotation? The issue has never really been decided but from that time onwards the Highlands of Scotland by and large have become sheep walks on the better moors, deer "forests" on the three or four million acres of the less accessible. Neither is the right answer at the present day.

When the Land Utilisation Survey of Britain carried out its work of field to field survey, described in the next chapter, in 1930-31 it thus found in places, though not complete over the whole country, material for comparison with the past. There were the maps and reports of the late 18th century, there were the details on the First Edition of the Ordnance Survey maps, the descriptions of Cobbett, the Tithe maps of the 1845 period, and the Ordnance Survey log books of the 1870 period. Over the century from 1845 to the present the comparison can often be made on a field to field basis. It is possible to draw a number of general conclusions from the comparison. The broad features of the land use pattern were firmly established long ago and on the whole there is remarkably little change.

There is stability of land use on the poorest lands. The mountain moorland and the coarse sandy heaths which defied reclamation in the past, still defy efforts at reclamation or improvement today. The absence of change is not for want of trying. A good example is the

M.L.—Q

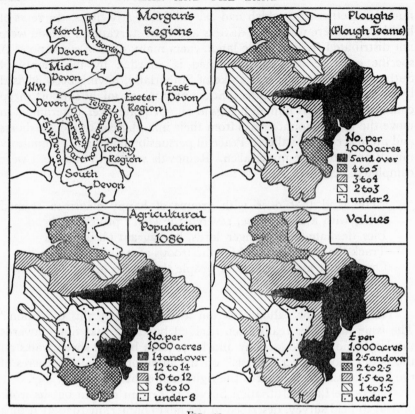

FIG. 43

Devon at the time of the Domesday Survey, 1086

The relative qualities and values of land had already been established (*after F. W. Morgan,* who delineated the physical and land use regions shown in the upper map).

attempt to tame Exmoor. After the expenditure of vast sums of money, the moor has won.*

Charles Vancouver in 1808 wrote of the need for seriously dealing with the reclamation of Dartmoor. But Dartmoor remains with very much its old boundaries. In 1942-45 some commons which perhaps had never before known the plough produced their quota of rye for the national loaf. But, for one reason or another, but especially where there are water difficulties, they are back where they were. This is

*C.S. Orwin : *The Reclamation of Exmoor Forest,* Oxford, 1929.

FIG. 44
The Soils of Devon as recorded by Charles Vancouver 1808

not, of course, to say that the improvement of submarginal lands is impossible : far from it. But I do suggest that it proves the strength of the opposition—the inherent quality of land.

At the other end of the scale there is remarkable stability of land use on the best lands. The loamy river terraces were sought out by the Anglo-Saxons ; they remain our best ploughlands. Despite all the economic vicissitudes of the centuries, the booms and depressions, the wide fluctuations in farm prices, the best lands have remained under continuous cultivation. The same is true of the best grasslands—such as Romney Marsh—or of lands which with careful management have been transferred from the best grass to the best arable, as in the Holland Division of Lincolnshire.

The maximum of change is on land of intermediate quality. There the economic factors have exercised their full effect. Lands inherently of only moderate quality (Categories 5 and 6 of the national scheme described in the next chapter) are abandoned in whole or part when agricultural prices are low, to be taken in hand again when prices rise. Particularly interesting are the good but heavy clay lands (Category

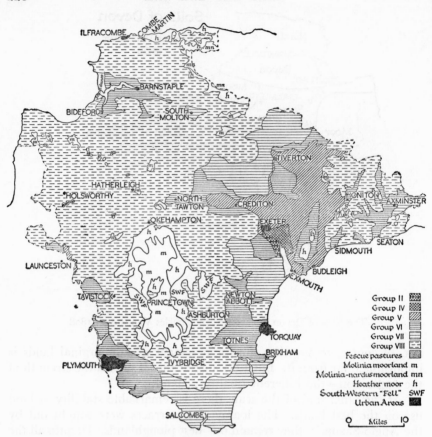

FIG. 45

The Grasslands of Devon as surveyed in 1940-41 by Sir George Stapledon and Dr. William Davies. The close agreement with land values in 1086 and soils should be noted

4). Not easy to manage, ploughable only when moisture conditions are just right and so workable only at certain seasons, they will yield the heaviest of all wheat crops when adequate labour is available. But they grow good average quality grass, rarely drying out and are more cheaply managed under grass. So from arable in the Middle Ages such clay lands came almost entirely under permanent grass in the latter half of last century and until 1938. On such lands the plough entirely altered the landscape during the last war.

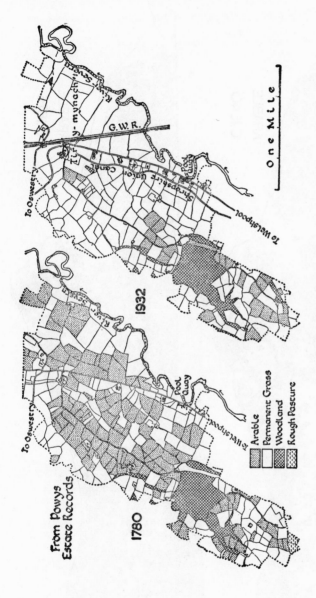

FIGS. 46 (*left*) *and* 47 (*right*)

A pair of maps of a Welsh estate showing changes in field boundaries and land use over 150 years. The field pattern has remained remarkably stable, despite such changes as the building of a railway. The land has remained improved farm land but with a lessened emphasis on arable. This is typical of most farming areas of Britain

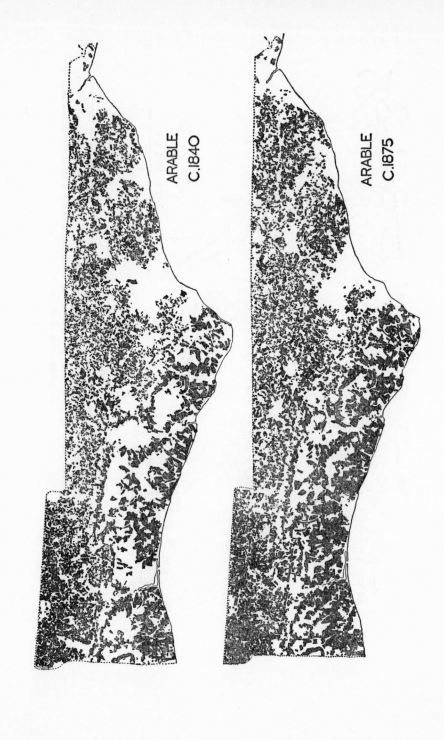

ARABLE
C.1840

ARABLE
C.1875

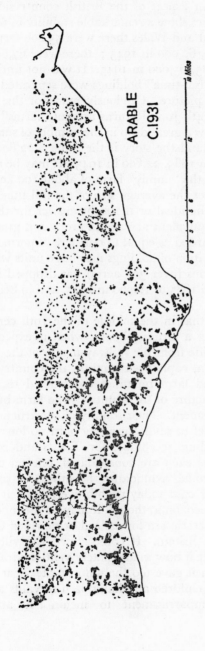

ARABLE
C.1931

0 1 2 3 4 5 6   12   18 Miles

FIGS. 48, 49 (facing) and 50 (above)

Three maps of southern Sussex, to show land use changes. The arable land has varied in total amount but occupies roughly the same areas at each period shown

There is an important aspect of the British countryside which is often overlooked. Statistics show a remarkable stability in number and size of farms. In England and Wales there were 55,000 farms between 50 and 100 acres in 1885, 60,000 in 1945 ; there were 67,000 between 100 and 300 acres in 1885, 65,000 in 1945. It was not until the Farm Survey of 1940-41 that "full-time" holdings were separated from part-time and gave the true picture. Obviously many of the "farms" of less than 20 acres in the 4th June Returns are not "farms" at all, and so the statistics which show a great drop in the number of small holdings are misleading. In Scotland the story is the same : 22,800 farms between 50 and 100 acres in 1885, 22,600 in 1945. It may be argued that this stability proved that the "family" farm of about 100 acres of crops and grass (which is in fact the average size of the full-time holding in England and Wales) is the ideal or the right answer for this country. Incidentally to get an equivalent acreage of crops and grass, 10 acres of rough grazing are regarded as equal to one acre of permanent grass. Those who urge consolidation of holdings into large units will find that the number of large farms (over 300 acres) has dropped in England and Wales from 16,600 in 1885 to 12,360 in 1945, and in Scotland from 2,766 to 2,368.

What is forgotten is that in the middle of the 19th century when the country was enjoying a farming boom, when unexpected capital flowed into the countryside from the railway mania of the 'forties and payment of compensation, capital was sunk in the countryside in the form of farm houses and farm buildings constructed to last. They crystallized the farm structure of the country. The farm buildings are often today an embarrassment. Their solidity of construction prevents their ready adaptation, let us say from a hay barn no longer required to a modern dairy steading acceptable for T.T. standards. If two farms of a hundred acres each are consolidated, one set of buildings and one farmhouse become redundant, the other set probably inadequate. With building costs today prohibitively high, it is easier to retain the farm unit, uneconomic though it may be for modern needs.

In other ways the Victoria era formalized or petrified or whatever is the right word, rural Britain. It consolidated the village by the addition of a solidly built if now inadequately ventilated and perhaps redundant school. Often it gave too the unfortunate vicar for himself and his wife, his dozen children and half-dozen servants a parsonage which today is an embarrassment to incumbent and Church Commissioners alike.

A good example of the complexity of the present position is afforded by the widespread problem of rural depopulation. Broadly speaking all purely rural areas have been losing population for many decades. This is far from being the same thing as agricultural decadence. With technological progress the same area of land can produce more than previously with a smaller labour force : improved mobility with every farmer a car owner as well as a tractor owner, concentrates servicing in the nearby towns. The craftsmen of old are no longer needed in the villages. To simplify a study of the problem I suggested a division of rural populations into three groups :

(a) primary rural—farmers and farm workers

(b) secondary rural—existing to service the first

(c) adventitious—the population living in the country by choice.

The first post-war Government policy was to restrict building in the country to farm workers' cottages and farms—to prevent the loss of agricultural land caused by adventitious building. I then demonstrated (see *The Land of Britain : Its Use and Misuse*, page 448) that the effect would be so to reduce the number of children that no average village would henceforth provide the minimum number of children which the Ministry of Education demanded to keep open the village school. Consequently, to maintain the social structure of the rural districts an "adventitious" population is now officially encouraged.

Enough has been said in this chapter to show that many features of our countryside which had been evolving through the centuries were fixed or consolidated in the form in which we know them during the the 19th century—not always to the advantage of the present day.

CHAPTER 17

TOWARDS NATIONAL LAND PLANNING
1895-1947

THE VICTORIAN ERA had been one of industrial and commercial expansion, of free trade and *laissez faire*. So far as the land of Britain was concerned, the towns continued to spread and sprawl over the countryside : an even larger proportion of the population became town dwellers, more and more of the children grew up in ignorance of all matters rural. From the peak of high farming in the 'sixties and early 'seventies, the influx of cheap food from overseas caused a downward trend in British agriculture, destined to last almost until the Second World War. Grassland and stock demand less labour than arable and crops and to economize in wages is the farmer's readiest, often his only, means of cutting down expenses. Consequently there was a steady swing over from ploughland to permanent grass throughout England and Wales and to a less extent in Scotland. At the same time the total acreage of improved land—crops and grass—steadily decreased not only because of the inroads of urban and industrial development but also because of the abandonment of marginal or poor lands. Land which it paid to plough and crop in what Cobbett called "dear corn times" was no longer profitable. It was allowed to "tumble to grass"—that delightfully expressive farmer's phrase which needs no explanation—and then to be invaded by the rough herbage of the neighbouring moors—reeds and rushes, coarse grasses, such as *Nardus* and *Molinia,* heath and heather, bracken and gorse. So the sum total of "rough grazings" expanded at the expense of improved farm land.

If anyone wanted farm land for building, or other forms of "development" there was no restriction. The power of the purse overcame all. There continued of course that strange phenomenon of all times in Britain : the urge to escape from towns to country quiet. Those who had made money in industry, trade or commerce built homes for their later years in the country : the richer laid out new estates and built mansions of varying pretentiousness and hideousness. Unlike those of earlier generations, the new-rich had forgotten country ways and knew not the land. They laid out their estates irrespective of land quality : they were no longer concerned that food should be home-produced. A map of some areas showing the later Victorian estates suggests the phrase, "You want the best lands, we have them." Although there were notable exceptions, appreciation of our heritage from the past was overshadowed by self-satisfaction with the present. Though this attitude was also characteristic of the Edwardians, the beginnings of a change were discernable.

The year 1895 has been put at the head of this chapter because it was in that year that there was founded, largely through the enthusiasm and drive of one devoted Englishwoman, Miss Octavia Hill, and two other practical idealists, Sir Robert Hunter and Canon Rawnsley, the National Trust for Places of Historic Interest or Natural Beauty. Contrary to widely held belief the Trust had no government or official backing : its object was to collect funds by private subscription and so to purchase and manage on behalf of the people of Britain historic buildings and so to save them from destruction. The addition of areas of natural beauty came somewhat later. The Trust has grown to be one of the largest property owners in the country and by 1963 had under its control in England and Wales over a thousand properties, covering more than 300,000 acres.

The latter part of Victoria's reign also witnessed a movement which has had a profound influence on the scenery of much of Britain. It was the move for better housing with adequate garden space. The worst type of back-to-back housing of the industrial towns could squeeze 128 dwellings or, to use a modern convenient if ugly phrase, "accommodation units" with access roads on to one acre of land. Allowing for the large families of the time, the residential density was of the order of 700 to 800 persons per acre. The position became a little better under the so-called "bye-law" housing, but the way to the future was pointed by such outstanding pioneer industrialists as William H. Lever (afterwards Lord Leverhulme) who established Port Sunlight in 1888-1889 and George Cadbury who laid the foundation of Bournville (1879) and the "factory in a garden" idea. Private enterprise launched Hampstead Garden Suburb in 1906 and it is interesting to recall that the early advertisements for this favoured retreat of the black-coated workers in the upper-income brackets portrayed a workman with his bag of tools slung over his shoulder trudging home to his cottage and garden. Ebenezer Howard published his scheme for a Garden City in *Tomorrow* in 1898 and the first Garden City, carrying out his ideas of a self-contained industrial-residential unit with adequate living space, was initiated by private enterprise at Letchworth in 1903, followed in 1920 by Welwyn Garden City. A housing density of ten units per acre, or 35 persons per acre, accepted for garden city standards, means that the same population occupies, for residential purposes alone, at least twenty times the area it did in the back-to-back slums. Despite vigorous advocacy by the Garden Cities Association (later renamed the Town and Country Planning Association), Howard's ideas were slow to gain recognition until adopted in general for the new towns of the post-Second World War. Instead, existing towns and cities throughout the country undertook great housing estates and dormitory suburbs or allowed private enterprise similarly to extend an unplanned sprawl. Speculative builders bought up a farm, erected a few houses for sale whilst allowing the remainder of the farm land to fall into disuse as building plots awaiting development. Another developer would then buy another farm farther out and a sort of leap-frog effect resulted. Naturally it was cheapest to build along existing highways so that ribbon development took place almost automatically along all the approach roads to towns. All this unplanned sprawl produced a crop of new problems—the provision of such essential services as electricity, gas, piped water and main drainage, as well as public transport to minimize the journey to work or access to shopping and recreational centres—and the growth of town planning became inevitable.

At first it was town planning—town and country planning was to come later. The viewpoint was purely urban, the country existed simply as a background to the developing town. Even under the 1932 Town Planning Act which made planning general though not compulsory throughout England and Wales there was still no thought of conserving agricultural land unless it appealed to the townsman as of special scenic interest. The concept of the Green Belt, limiting the outward growth of towns, was likewise an urban one. Briefly and ephemerally the submarine menace of the First World War had focussed a spotlight on the importance of home food production. The plough-up campaign of 1918 proved but a minor interruption of the general decline of home agriculture in general and home crop production in particular.

Nevertheless the appreciation of the precious heritage of rural Britain was growing. For obvious reasons the Commons, Open Spaces and Footpaths Preservation Society had been founded as long ago as 1865, when enclosures had all but eliminated the commons in many counties, but the broader concept of conserving the best from the past was foremost in the establishment of the Council for the Preservation of Rural England (C.P.R.E.) in 1926, followed by a similar Association for Scotland in 1927, and the Council for the Preservation of Rural Wales in 1928. Concern for wild life had led to an agitation for Nature Reserves—the Society for Promotion of Nature Reserves was founded in 1912.

Amongst professional bodies it is interesting to recall the establishment of the Town Planning Institute in 1914, under the joint auspices of professional architects, surveyors and land agents. For very many years a technical training in one of these three professions was an essential requirement for membership of the Institute : the genesis of town planning was obvious. Now the planner is concerned with an overall view of the whole land, its resources and their orderly development and effective use, and those with other basic trainings are admitted as candidates for membership.

Side by side with preservation of past heritage, creation of new beauty was not forgotten. The great days of Capability Brown and the landscape gardeners of the 18th Century were doubtless in the minds of those who founded the Institute of Landscape Architects in 1929 (compare Plate XXXII). That modern transport need not scar the landscape is suggested by the Roads Beautifying Association (1928).

However useful these many voluntary and professional bodies may have been and still are in moulding and directing public opinion, it was the establishment of certain somewhat glaring trends in the years between the two World Wars which led to government intervention.

The 1931 Census first demonstrated clearly what had been suspected for some time. The great growth of population and associated industrial and urban expansion were concentrated in two main areas in Britain—Greater London and Greater Birmingham. In more general terms it was within a rectangular tract stretching from Liverpool and Leeds on the northern side to Southampton and the south-east coast on the southern. Although often referred to as a southern drift of industry it was not really a movement in the sense that phrase suggests. Industries did not move, but as they died in the older industrial areas they were not replaced : new factories were established in new areas. There grew up the spectacle of the older industrial areas of Scotland, the north-east (Northumberland and Durham), the north-west (West Cumberland) and South Wales with dead and dying towns and much unemployment. The opprobrium of the term "depressed areas" was later replaced by "special areas" and still later by "development areas" but in the inter-war years depressed they certainly were. It was in 1938 that the Government set up the Royal Commission on the Geographical Distribution of the Industrial Population under the Chairmanship of Sir Montague Barlow. Broadly the Commission found that the time was ripe for a deliberately planned location of industry to secure its wider dispersal and distribution. Before the Commission reported, however, Hitler's bombs had already done much to disperse industry.

The outbreak of the Second World War brought the British farmer into the front line of defence. It was soon apparent that he alone could save the country from defeat by starvation. The uncertainty which followed the Munich "agreement" of 1938 gave time for the perfecting of plans to be put into operation should war break out so that when war did come in September, 1939, the agricultural war machine was slammed into top gear overnight. Ploughland is not only more productive than permanent grassland : it is more flexible. Human beings can consume direct not only the cereals grown as crops—alternatively they serve as animal fodder—but also the root crops such as turnips, swedes and potatoes, normally grown in the main as winter feed for animals. So the Cabinet set a target for home food production which, with reduced imports, was to make up the national rations. In its turn the Ministry of Agriculture or the Department of Agriculture for Scotland allotted to each county its appropriate share of production. In each county the County War Agricultural Executive Committee (C.W.A.E.C.) comprised largely of local farmers, land owners and land agents decreed what each farmer must do, and had power of compulsion, even dispossession, if he did not. So it came about that in the first four years of war, from 1939 to 1943, despite a reduced male labour force, the substitution of 80,000 members of the town-recruited Women's Land Army, shortages of all sorts, Britain became the most highly mechanized farming country in the world. and the home production of many commodities was doubled. Total home output increased in sum from about 35 to 55 per cent. of total consumption. Imports of animal feeding stuffs dropped to about a tenth of the pre-war figure and, broadly, each farm became a self-contained unit, producing what it needed for the support of its own livestock. There was concentration on milk and the maintenance of priority milk supplies for mothers and children had much to do with the excellent health record during the War. Beef cattle, sheep and pigs were reduced in numbers : much unsuitable land was pressed into service, at least temporarily, for cereal production.

What were the effects of the wartime revolution on the scenery and the land of Britain ? A rather unkempt and neglected countryside gave place to a neat and trim one, consciously proud of the job it was doing and wearing an air of efficiency. Largely with the use of prisoner-of-war labour a huge backlog of hedging, ditching, and draining was wiped off. Hedges were properly trimmed and laid, often for the first time in years, ditches were cleaned out, reedy pastures drained. Ploughed fields appeared where for years there had been only grass— a change particularly noticeable in the Midland "grassy shires" such as Northampton and Leicester—and figures show that the plough-acreage climbed back almost to its total of the high farming days in the early seventies. The horse gave place to the tractor, the hay wagon to the motor truck. There was much use of common land—some ploughed and cropped for the first recorded time in history—and improvement of hill lands was pushed back to long forgotten upper limits.

To a large extent the position was consolidated by post-war legis-lation, especially the Agriculture Act of 1947. Mechanization has come to stay and so undoubtedly has a greater reliance on leys rather than permanent grass. Many farmers who previously never used a plough have learnt to "take the plough round the farm" and periodically to break and reseed their grasslands. Much of the land of Britain emerged from the war in better heart than ever before, and though common lands and marginal lands in some areas have been going back to what they were, many wartime changes are destined to remain.

Even in the darkest days of the war, Britain showed the nation's faith in the future by making post-war plans. Comprehensive land planning was indeed born during the war. In April 1941 Lord Reith, the first Minister of the newly created Ministry of Works and Buildings (afterwards Works and Planning) appointed a Consultative Panel one of whose committees supervised the early work of the Research Maps Office responsible for the National Planning Series of Maps on the scale of 1 : 625,000. Shortly afterwards Lord Reith in consultation with Mr. R. S. Hudson (afterwards Lord Hudson) the Minister of

Agriculture set up the interdepartmental Committee on Land Utilisation in Rural Areas under the Chairmanship of the late Lord Justice Scott. I had the honour to serve as Vice-Chairman of that typical jury of twelve British citizens drawn from many walks of life, and we worked unceasingly for the year before our Report appeared in August, 1942. Under our terms of reference we were required to "consider the conditions which should govern building and other constructional development in country areas consistent with the maintenance of agriculture .... having regard to .... the well-being of rural communities and the preservation of rural amenities." We interpreted our duties liberally and our Report, hailed widely as a charter for the countryside and a blue print for the future, was an immediate best seller. We were not concerned with financial problems : they were being considered simultaneously by the Expert Committee on Compensation and Betterment being presided over by Mr. Justice Uthwatt. The Scott Report included scores of recommendations, large and small : in due course nearly all have been adopted by successive governments and embodied in legislation. Our pleas for a Minister of Planning to co-ordinate the planning work of other departments and aided by a small commission but without departmental responsibilities went, however, unheeded. Instead the Government of the day set up (in 1943) a separate Ministry of Town and Country Planning and, as we had anticipated, it proved difficult to separate the Minister's co-ordinating functions from the planning for which he was directly responsible.

Many of our recommendations were embodied in the Town and Country Planning Act of 1947. Others gave rise to a whole crop of further committees before they were accepted. Our recommendations regarding National Parks gave rise to the investigation of the problem by the late John Dower and the Dower Report, and then to the Committee on National Parks presided over by Sir Arthur Hobhouse and so finally to the establishment of National Parks. Our recommendation on nature reserves led eventually to the formation of the Nature Conservancy (incorporated by Royal Charter, 1949) ; our recommendation on footpaths and access eventually to the provisions included in the National Parks Act.

It is worth while to look briefly at some of the background to the Scott Report. In the years between the wars—between 1927 and 1939 —there was an annual net loss of open land of some 60,000 acres to industry, housing and other constructional development. Naturally the bulk of the loss fell on improved farm land, since towns can scarcely be built on mountain moorland, and much indeed on the best agricultural lands of the country. Land which is level or gently undulating and well drained is not only good agricultural land but is also eminently suitable for the horizontal layout of a modern factory and even if it does not form the most attractive housing sites aesthetically, it is the less expensive to develop*. In the short space of those 12 or 13 years the total loss of agricultural land had been more than equal to the combined areas of Hertfordshire and Bedfordshire. Nothing can alter the fact that the total area of Great Britain, some 56,000,000 acres, must supply all the land needs of 51,000,000 people. The position is far worse if one takes England alone with 32,000,000 acres for 46,072,000 people (1961 census). In Lowland Britain we have less than three-quarters of an acre per head of land of all sorts, or only a little over half-an-acre of improved farm land.

Land planning has been forced on Britain not by reason of pressure from any particular political group or groups but by the sheer necessity of using to the best advantage the extremely restricted resources of land.

Even in such a long settle and one would think well-known land there are some surprising gaps in the factual knowledge essential to the planning of land use. There are two approaches to physical planning. One is to record the present position—this means to record existing land use; then to seek to determine the reasons for that position and to trace the respective and relative influence of the physical, historical, social and economic factors. Many, even most, of such factors are still operative and determine trends in development. Planning becomes the work either of deliberately countering existing trends or of encouraging them. It is true that, especially in the underdeveloped countries of the world, there are those who prefer a second approach, a short cut. They would ignore the present position and the factors which have led to it and would determine land potential.

It was with the primary object of remedying the first defect—lack of knowledge of existing land use—that I organized in 1930 the Land Utilisation Survey of Britain.† The story of the survey has been elsewhere: our immediate task was to record on the 22,000 separate published sheets of the 6-inch to one mile (1 : 10,560) Ordnance Survey maps the then-existing use of every field and piece of land. The work was organized on a county basis throughout

* See Plate XXXI.
† A second and more elaborate land-use survey is now in progress under the direction of Miss Alice Coleman, King's College, London, Maps are being published on the scale of 1 : 25,000.

M.L.—R

England, Wales and Scotland, and the Isle of Man. Speed was important as a snapshot picture was needed ; the bulk of the field work was in fact carried out in 1931 and 1932. Nearly all was done as a voluntary exercise by the schools, colleges and universities of the country, and perhaps a quarter of a million young people helped in the task. The 6-inch field sheets were checked, edited and revised where necessary and the work reduced to the scale of one inch to one mile. The publication of the one-inch sheets began with the issue of the first two on January 1st, 1933, and continued steadily until eventually 170 maps were published covering the whole of England and Wales, the Isle of Man, and the more populous parts of Scotland. Unfortunately the main stock and the printing plates of the maps issued before 1941 were destroyed in an enemy raid, so that the earlier sheets are out of print and rare. Later a generalized map on the scale of 1 : 625,000 (approximately 10 miles to one inch) in two sheets was published in the National Planning Series. The results of the work were analysed in a series of county reports in 92 parts or nine quarto volumes under the title of *The Land of Britain*. The general results are summarized in my book *The Land of Britain : Its Use and Misuse*, (Longmans, 1948, 3rd Edition 1962).

The maps of the Land Utilisation Survey refer to a period when British farming had reached its lowest inter-war ebb. The maps mirror the official agricultural statistics but, because the areas are accurately delineated in maps, permit the influence of such factors as relief, elevation, soil, local climate and accessibility to be assessed. The maps naturally form a standard of comparison with the past and can now be so used in a study of changes resulting from wartime stimuli and post-war conditions.

It may well be argued that existing land use at any one time is no real key to land potential or inherent quality. In a long settled country such as Britain, land use studied historically (see above page 225) does in fact afford a close guide to land potential though far from giving the whole answer.

It was when the Government awoke to the very rapid loss of good agricultural land that the famous declaration of policy was made in both Houses of Parliament that henceforth the Government would "seek to avoid the use of good agricultural land for housing development where other and less valued land could be appropriately used." There were two difficulties in the application of this principle. There was no accepted definition of "good agricultural land" and certainly

Good Quality Land ▓

Good Quality Land
mixed with poorer
quality land ⬚

0   Scale of Miles   100

FIG. 51
The good Quality Farm Land of Britain
Categories 1, 2, 3 and 4 of The Land Utilisation Survey

no attempt had been made to map its distribution. The nearest approximation to maps of types of land were in the Agricultural Reports of 1793-1800.

It was Sir Montague Barlow as Chairman of the Royal Commission who asked me as Director of The Land Utilisation Survey whether it would be possible to draw up a scheme of land classification and to map the distribution of the different types. As a result of much discussion with experts in many fields a preliminary statement was published in 1941; a small outline map and revised scheme were published in the Scott Report (1942). After conferences with the Soil Survey of England and Wales in 1943 an agreed scheme of ten types, falling into the three major categories of Good, Medium and Poor, was agreed and adopted by the Ministry of Agriculture (Circular R.L.U.6 of 1943) and the Ministry of Town and Country Planning. Maps were prepared in manuscript on the scale of one inch to one mile. In 1944 a general map on the scale of 1 : 625,000 was published as one of the National Planning Series by the Ministry of Town and Country Planning showing the distribution of the ten types of land.

The types of land are defined as follows :

MAJOR CATEGORY I—GOOD QUALITY LAND

Highly productive when under good management. Land in this category has the following characteristics :

Site 1 not too elevated ;
  „   2 level, gently sloping or undulating ;
  „   3 favourable aspect.

Soil 1 deep ;
  „   2 favourable water conditions (actual or potential) ;
  „   3 texture, mostly loams but including some peats, sands, silts and clays.
The three criteria under site apply to all the four types of good agricultural land:—

1. *First Class Land* capable of intensive cultivation, especially of foodstuffs for direct human consumption. The soils are deep and in texture are mainly loams but include some peats, fine sands, silts and loamy clays. Drainage must be free but not excessive and the soils must not be excessively stony, and must work easily at all seasons.

2. *Good General Purposes Farmland.* This land is similar to the first but is marred by (*a*) less depth of soil, or (*b*) presence of stones, or (*c*) occasional liability to drought or wetness, or (*d*) some limitation of seasons when the soil works easily resulting in a restriction of the range of usefulness. When the conditions are such that the land is particularly suitable for arable cultivation the designation 2(A) may be used ; when the conditions are such—notably in wetter regions—that sown

grasses or permanent grassland are particularly suitable the designation 2(AG) may be used and such land has been shown by a separate colour on the map.

3. *First Class Land* with water conditions especially favouring grass. This land is similar to 1 but as a result of (*a*) a high permanent water-table, or (*b*) liability to winter or occasional flooding, or (*c*) somewhat heavier or less tractable soils it is unsuitable or less suitable for arable cultivation than for grass. Such land may be converted into Category 1 by drainage or prevention of flooding but this is a major operation.

4. *Good but Heavy Land.* Although such land has soils of good depth and the natural fertility is often high, the soils are heavy—mostly of the better clays and heavy loams—with the result that both the period of working and the range of possible crops are restricted. Because of these difficulties, most of this land in England, though not in Scotland, was down to grass before the Second World War. Some would place such land in the category of Medium Quality Land, yet for certain crops such land is the finest of all. In England, because of the predominance of grass, the designation 4 (G) may be used, but 4(A) would be more appropriate on the boulder clays of Scotland.

MAJOR CATEGORY II—MEDIUM QUALITY LAND

This is land of only medium productivity even when under good management. Productivity is limited by reason of the unfavourable operation of one or more of the factors of site or soil character, *e..g*, by reason of

    Site 1 high elevation ;
      „   2 steepness ;
      „   3 unfavourable aspect.
    Soil 1 shallowness ;
      „   2 defective water conditions.

It is obvious that a wide range of conditions—indeed an almost endless combination of one, two or more deleterious factors—is included in this major category. The Land Utilisation Survey recognised that the two chief types of harmful conditions were (*a*) soil and (*b*) relief, and though a wide range is possible within each, this gives Categories 5 and 6.

5. *Medium Quality Light Land.* This is land defective by reason of lightness and, usually, shallowness of soil. The moderate elevation, relatively gentle slopes and consequent aspects are all satisfactory. There are several distinct types included within the category the chief being :

(*a*) shallow light soils on chalk or Jurassic limestones—the downland and Cotswold soils where ploughable 5(A) ; where not ploughable 5(G).

(*b*) shallow soils on some of the older limestones and shallow light soils which occasionally occur on other older rocks—usually not ploughable owing to rock outcrops, hence 5(G).

(*c*) light soils, including gravels, not necessarily shallow, which occur on solid formations such as the Bunter Sand of Sherwood Forest or on superficial deposits as in Breckland or on consolidated sand dune areas as in some of the coastal districts of Scotland. Such land is usually best managed as ploughland, hence 5(A).

6. *Medium Quality General Purpose Farmland.* This is land defective primarily by reason of relief—land broken up by steep slopes, with patches of considerable elevation, varied aspect and varied water conditions. In consequence, soils are varied, often deficient by reason of stoniness, shallowness, heaviness or in other ways. When a tract of country of this general character is studied in detail it is usually possible to resolve it into a mosaic of small tracts or patches, it may be only a part of a field in size, of land varying from Categories 1 to 10. Most land of Category 6 is usually equally suitable for crops or grass, hence the designation 6 (AG).

MAJOR CATEGORY III—POOR QUALITY LAND

Land of low productivity by the extreme operation of one or more factors of site and soil.
- (*a*) extreme heaviness and/or wetness of soil giving poor quality heavy land or land in need of extensive drainage works ;
- (*b*) extreme elevation and/or ruggedness and/or shallowness of soil giving mountain moorland conditions ;
- (*c*) extreme lightness of soil with attendant drought and poverty giving poor quality light land ;
- (*d*) several factors combining to such an extent as to render the land agriculturally useless or almost so—such as shingle beaches or moving sand dunes.

7. *Poor Quality Heavy Land.* This includes the more intractable clay lands and low-lying areas needing extensive drainage works before they could be rendered agriculturally useful. For convenience, undrained mosses have been included though the soils they might eventually yield would not necessarily be heavy. The heavy clay lands tend to be in grass hence the designation 7(G).

8. *Poor Quality Mountain and Moorland.* The wide variety of land included in this category is apparent from the varied character of the natural or semi-natural vegetation by which it is clothed.

9. *Poor Quality Light Land.* This category includes the so-called "hungry" or over-drained lands, usually overlying coarse sands or porous gravels and hence including both coastal sand dunes and the inland sandy "wastes" or heathlands.

10. *Poorest Land.* In its present state this land may be agriculturally useless, but this is not to deny possibilities of reclamation. Salt marshes can be drained, sand dunes fixed, and so on.

Special interest attaches to the following table, based on a careful county by county calculation made by the Ministry of Town and

Hunting Aerosurveys Ltd.

Plate XXIX.—THE NEW LANDSCAPE IN THE MIDLANDS

This is an opencast iron ore working, near Corby, Northamptonshire. The " hill and dale " created by the system of working is well seen in this air view. The dark area is a plantation of spruce—one of the most effective re-uses of the land

Plate XXXa.   Aerofilms Library

b.   Fox Photos Limited

Country Planning* (Research Maps Office, under Dr. E. C. Willatts) of the areas involved.

CLASSIFICATION OF LAND IN BRITAIN

	England and Wales†		Scotland		Great Britain	
	Acres	%	Acres	%	Acres	%
Category I—Good	17,845,900	47.9	3,963,300	20.8	21,809,200	38.7
1 First Class	1,963,100	5.3	396,800	2.1	2,359,900	4.2
2 Good General farmland						
2 (A) for ploughing ..	7,065,600	18.9	1,735,900	9.1	8,801,500	15.6
2 (AG) crops or grass ..	2,636,900	7.1	192,900	1.0	2,829,800	5.0
3 First Class, restricted ..	1,234,800	3.3	8,700	0.0	1,243,500	2.1
4 Good but heavy ..	4,945,500	13.3	1,629,000	8.6	6,574,500	11.7
Category II—Medium ..	11,933,800	32.0	2,877,400	15.1	14,811,200	26.3
5 Medium light land						
5 (A) for ploughing ..	2,402,100	6.4	77,400	0.4	2,479,500	4.4
5 (G) not for ploughing	220,300	0.6	300	0.0	220,600	0.4
6 Medium general farmland	9,311,400	25.0	2,779,700	14.7	12,111,100	21.5
Category III—Poor.. ..	6,350,900	17.0	12,113,800	63.5	18,464,700	32.8
7 Heavy land	825,900	2.2	54,100	0.3	880,000	1.6
8 Mountain and moor ..	4,516,800	12.1	12,001,700	62.9	16,518,500	29.3
9 Light land	811,800	2.2	57,900	0.3	869,700	1.5
10 Poorest land	196,400	0.5	100	0.0	196,500	0.4
Closely built over	1,142,700	3.1	114,200	0.6	1,256,900	2.2
Total	37,273,300		19,068,700		56,342,000	

† Including the Isle of Man.

* Now merged in the Ministry of Housing and Local Government.

Plate XXXa. THE CONSTRUCTION OF WESTERN AVENUE, LONDON, IN THE INTER-WAR YEARS
A typical example of a new highway slashing across the country, ignoring fields, farms and the older settlements.

b. THE GREAT WEST ROAD OUT OF LONDON
Inadequate even before completed : trees have been planted to make an avenue and a small green planted with trees and shrubs. The waste ground is "valuable building land" lying idle.

It will be noticed that only 4 per cent. of the surface of Britain is put into the Category I of First Class—in all less than 2½ million acres.

Details of the criteria used in deciding the category into which a given tract of land falls are given in Chapter XVII of my book *The Land of Britain*. Critics urge that too much attention is given to present and past use, not enough to inherent qualities. The truth is that in the absence of detailed soil surveys over much of the country information based on a detailed knowledge of soil is lacking. Thus the second approach to land classification, that of land potential, cannot be attempted in Britain yet. The Soil Survey of England and Wales and the Soil Survey of Scotland are at work to remedy this serious gap in knowledge.

During the war there was much destruction of our towns by aerial bombardment. Studies were undertaken to serve as a basis for reconstruction and many elaborate "Plans" were prepared. In the majority of these the new approach to land resources is at once apparent. Conscious efforts are made to conserve the better agricultural lands, to consider the importance of balanced or economic farm units and to dovetail the interests of food production and amenity. Proposals for new towns took note of the distribution of land types and, generally speaking, efforts were made to avoid using the best agricultural lands. I contributed chapters on agriculture and land classification to the Plans by Sir Patrick Abercrombie and others on Greater London, Merseyside, Plymouth, Hull, Bath, Warwick and elsewhere. The proposals for new towns around London afford a good example of the attempts to use poorer land though not always with success. With the end of the war, the recession of the danger of starvation, public memory is short and the farmer at least fears his significance in the national life is again in danger of being forgotten. But we can never go back to the old profligate attitude towards our land resources. Land planning is here to stay and so the landscape of Britain is destined to be moulded perhaps as never before by the hand of man. Of course national planning in Britain is nothing new. It was put into practice by the Romans—otherwise we should not have any straight roads.

CHAPTER 18

THE SECOND ELIZABETHAN AGE
AND THE FUTURE

IT HAS very rightly been said that any nation has only two ultimate assets—its land and its people. Britain has a small land and many people. Yet that small land is one of infinite variety, rich in scenic beauty, in heritages from the past: rich in its physical attributes—its mineral wealth, its soil, its climate—for present and future use. Some would say its people, topping the 51 million mark, are too many and should go elsewhere in large numbers. Actually they are increasing by a quarter of a million a year.

Undoubtedly the problem of the moment and for the future is how to use our strictly limited land resources to satisfy the many needs of our people and fulfil their legitimate desires. That is the purpose of planning—physical planning or town and country planning. Under the provisions of the Town and Country Planning Act of 1947 planning for the future has become compulsory and the unit for the purpose has been made the county—including of course the city or borough which has been given county status. Since the counties of Britain date for the most part from pre-Norman times they are today units which range widely in size, population and resources but a Commission set up a few years ago revealed such strong county loyalties and such bitter opposition to change that it was dissolved. Later proposals have not found easy acceptance.

So each county was set the task of surveying its area and preparing an orderly plan of development extending in the first instance over a period of 20 years which some optimist had designated the forseeable future. Not many of the plans were ready by mid-1952 (five years after the passing of the Act) for submission to the Minister, but a year later many were completed and submitted for approval by the central authority—now the Minister of Housing and Local Government.

It may of course be argued that a series of local plans however good does not add up to a national plan. There are, however, certain overriding principles which have been laid down or accepted by the central government and to that extent county plans fall into a general framework. Behind the whole there must be the need to satisfy the varied needs and legitimate aspirations of our people in so far as that is possible by the planned use of our small area of land.

In the first place the prosperity of this country depends to a major degree on industry. National policy is to secure a wide spatial distribution of industry, to encourage re-development in the older industrial areas, especially on the coalfields, and to restrict use of new land especially in such areas as Greater London, where population growth has been over rapid. Many industries are in fact fixed by physical factors—the extractive industries such as coalmining are obviously fixed by the occurrence of the minerals concerned. Where minerals occur in such abundance that only a fraction of total resources will ever be used, the localities of mines and quarries can be subject to other planning considerations. Not infrequently mineral extraction is in conflict with other uses of land, notably agriculture. The emphasis should not be on prevention of the use of important resources but on the restoration of land to other uses after the extraction of the mineral. Plates 24a, XXIX and XXXII deal with some aspects of the problem. Many heavy industries must be so located as to minimize transport costs. It is for this reason that large new iron towns have been established in what were formerly country areas at Corby and Scunthorpe because more than three-quarters of the weight of the iron ore extracted is waste and so the fuel and flux have been brought to the source of the iron ore rather than the reverse. Many industries need a location near tide water : other industries give rise to dust, fumes or smoke and must be located where least harm will be caused. The number of light industries which can freely be located at will is really very small.

A great demand for land is caused by the need for modern housing compatible with modern living standards. The British preference on the whole is for detached two-storeyed dwellings with separated plots of land. Semi-detached dwellings come next, lower down the scale are terrace houses. On the whole it is only young married people (couples

---

Plate XXXIa. THE VILLAGE OF EDGWARE IN 1926
The expansion of the village into a dormitory suburb of London has begun, though the tube railway has not yet been brought to the site cleared for it (*bottom right*).
(*Aerofilms Library*)

b. THE SUBURB OF EDGWARE IN 1949
The urban pattern has completely submerged the old rural pattern.
(*Aerofilms Library*)

By courtesy of G. A. Jellicoe

Plate XXXVII.—THE TWENTIETH-CENTURY LANDSCAPE ARCHITECT AT WORK

This is a photograph of a model to show proposed landscape treatment of Cement Works in the Hope Valley, Peak District. Trees are planted to hide the mouth of the limestone quarry, to screen the dusty works, along the approach roads and bordering the artificial lake which is the pit from which clay has been dug. A woodland covers the quarry-scarred hill on the right. This scheme, by G. A. Jellicoe, was commissioned and put in hand in 1946 by the public-spirited owners of the Cement Works (*pp. 237, 250*)

before the coming of a family) or old people who vote for flat life. If all our people now living in substandard, obsolete or obsolescent houses are to be rehoused at a density of 10 or 12 houses to the acre (35 to 40 people) the demand for land is very large. It is unfortunate that any discussion on desirable housing standards tends to engender heat and to range disputants on the side of the "garden city enthusiasts" or the side of the "flat-dwellers". Fortunately in much of the post-war housing moderate counsels have prevailed and a mixture of types of accommodation is being provided. Owing to the northern latitude of the islands and the low angle of the sun in winter, blocks of flats cannot be closely spaced if access to sun and air is to be adequate. A fierce controversy has raged around the subject of the productivity of gardens. Because of unsubstantiated statements that when land was converted from farming to a housing estate at open density the production from the land was increased because of the enthusiasm and skill of the gardeners, I undertook some preliminary surveys. The results are briefly recorded in *The Land of Britain : Its Use and Misuse*, page 190, and show that even with open density housing only 20 per cent. of the land remains available for cultivation. On this land lack of skill and marketing arrangements offset hand labour and enthusiasm so that output per unit area is no more than under the same crops on a commercial farm. On the other side the argument of retail money value of garden crops was produced and led to careful studies being undertaken by the Research Staff of the Agricultural Land Service (Ministry of Agriculture). It is interesting that garden production is highest on land easily worked throughout the year which is normally too poor to make good farm land and is highest on gardens of medium size, *i.e.* where the householder can work himself all he has*.

The need of land for recreation raises a number of problems at different levels. Numerous and conveniently placed parks and open spaces in all residential areas ; adequate playing fields attached to schools (but are 15 acres as the minimum standard for a new school too much ?), playing fields for young people and adults (6–7 acres per 1,000 inhabitants gives 300,000 acres as a total requirement), National parks, access areas, coastal strips and footpaths for vacation use. Nature reserves also come into the picture. The question becomes what can the nation afford ? How far can multiple use, as for recreation, grazing and water supply in moorland areas, be pressed ?

* See The Use of Gardens for Food Production, Mackintosh, P. and Wibberley. G. P., *Jour. Town Planning Inst.*, 38, 1953, 54-58.

Transport and communications have their land needs : the country undoubtedly should have at least a limited number of high speed motorways constructed as far as possible to avoid use of good farm land. Nevertheless with a central strip, two- or three-lane carriage ways, pull-offs, cycle tracks, pedestrian paths, grass verges and scenic margins the consumption of land could be enormous. Plate XXX suggests the difficulties to be faced. Civil airfields are large modern consumers of land.

The defence services demand large tracts for training and exercise and there are many miscellaneous demands—not least reservoirs for water supply and power works.

All these in a way are "land spenders". The "land savers" are agriculture and forestry. The old antagonism between these two has largely gone, though only in very recent times. Even at present each inhabitant of Britain only has half an acre of food producing land against the one acre needed to support him on present standards and present methods of farming, well applied.

We *must* use open farm land for other purposes but it should be done with our eyes open : all competing needs carefully weighed and considered. The broad general principle is that of optimum use— finding the most important use in the national interest of every inch of land. Above all is perhaps the complete elimination of waste land. Clearly too more attention ought to be paid to land reclamation—of coastal marshes for example, though the disastrous floods of 1952–3 have somewhat dampened enthusiasm in this direction.

From whatever point of view one regards the matter, the demands of different forms of development on what is at present open land, for the most part improved farm land, are serious. A few years ago I made a number of simple calculations of land which would be required in the near future.* In the first calculation I made the assumption that half our total population were due to be rehoused and that they would be rehoused at densities and standards being adopted for the new towns of Crawley and Harlow and that the total area taken up by the new housing areas was to be regarded as used for urban purposes. The result was the staggering total of 2,250,000 acres. This is equal to the total area of three average-sized counties and if the bulk of this were in England and were taken up over land at present in crops and grass it would take up nearly a tenth of the land so remaining at present.

*Stamp, L. D. (1950). Planning and Agriculture. *Jour. Town Planning Inst.*, 36, 1950, 141-152.

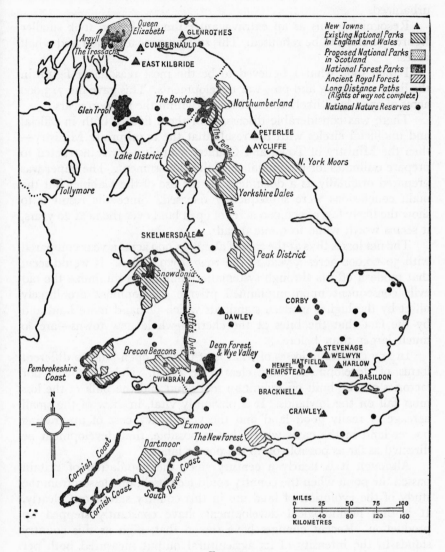

Legend on map:

New Towns ▲
Existing National Parks in England and Wales
Proposed National Parks in Scotland
National Forest Parks
Ancient Royal Forest
Long Distance Paths (Rights of way not complete)
National Nature Reserves ●

Map labels:

Queen Elizabeth
Argyll & The Trossachs
GLENROTHES
CUMBERNAULD
EAST KILBRIDE
The Borders
Northumberland
Glen Trool
The Pennine Way
PETERLEE
AYCLIFFE
Lake District
N. York Moors
Tollymore
Yorkshire Dales
SKELMERSDALE
Peak District
Snowdonia
Offa's Dyke
CORBY
DAWLEY
STEVENAGE
WELWYN
Dean Forest & Wye Valley
Brecon Beacons
HATFIELD
HARLOW
HEMEL HEMPSTEAD
BASILDON
Pembrokeshire Coast
CWMBRAN
BRACKNELL
N
CRAWLEY
Exmoor
The New Forest
Dartmoor
Cornish Coast
South Devon Coast
Cornish Coast

MILES
0   25   50   75   100
0   40   80   120   160
KILOMETRES

Fig. 52

National Parks, Forest Parks, Long Distance Paths, some National Nature Reserves, and new towns, 1963.

Something approaching a sixth of all Lowland Britain would then be urbanized.

Recognizing this as an extravagant estimate, I then took smaller figures—a third to be rehoused. This still gave me a million and a half acres.

I then took what I believed to be the most reasonable figure in the light of existing and post-war developments. This gave me 713,000 acres of open land likely to be absorbed within the next 20 years.

There was considerable discussion of these figures both in official and unofficial circles with the result that the appropriate Ministry— then the Ministry of Town and Country Planning—was instructed to prepare estimates for the guidance of the Government. The paper was prepared originally as a document for the use of the Cabinet but the main conclusions were subsequently released. Since the result is to show the likely loss of 700,000 acres of open land over the next 20 years, it seems worth while to quote details.

The net loss is thus at the rate of about 35,000 acres a year compared with 50–60,000 acres a year in the 1927-1939 period. It would seem that modern efforts through systematic planning to minimize the old evils consequent upon unplanned private development are largely offset by the higher modern standards which demand more land and by the fact that the bites at the cherry—whole new towns—are so much larger than before.

In any case large losses of what is at present open land to different forms of development must clearly be faced. Neither towns nor factories can be built on mountain moorland and the bulk of the loss must fall on the lowlands. It is obviously vital, in view of the small acreage of really good land and the very heavy cost of up-grading poorer land in the cases where that is possible, that development be directed as far as possible on to the poorer lands.

Although it is nearly a century since the population of Britain passed the point when the country could feed itself, one lesson from the study of the evolution of land use in this country stands out clearly. It is that technological developments have constantly stepped up production. Recent changes have shown that, high as this country stands in the intensity of its agricultural output measured both per acre and per worker, we are far from having reached the limits of what can be done if we are forced. "If we are forced"—the stimulus may come from a collapse in our overseas export markets, from a rise in world commodity prices, from world shortages of foodstuffs ;

unfortunately another "national emergency" cannot be ruled out. Whatever we may need at the moment, it ought never to be forgotten that once land is covered with bricks and mortar its restoration to food production is virtually ruled out.†

Two types of change in land use may be envisaged. One is the more intensive use of all types of land. For example, a third of the country is mountain moorland. Perhaps a third of this could be improved as pasture be reseeding, a third could be afforested, leaving only one-third as a "hard core" much as it is at present; space still where wild nature can reign. In this regard I see no reason to alter a table (p. 256), showing possible developments, which I published some years ago*.

Since the publication of the first edition of this book in 1955 changes have been rapid and numerous. Reference has already been made to some of the most conspicuous in the countryside—the present ubiquity of black and white Friesian cattle, the disappearance of the farm horses, the widespread appearance of fields of kale. The problem of common land has been studied by a Royal Commission and is the subject of a special volume in the *New Naturalist* Series. A far-reaching and subtle change is due to what I have called the chemicalization of agriculture. Increased yield of crops through use of balanced fertilizers is one thing but it is the use of sprays and dressings against pests and diseases which may upset the balance of nature in ways yet unknown. Certainly many wild flowers have disappeared from our fields and hedgerows and the work of the Nature Conservancy is now of vital importance.

Nor must we ever forget what is probably the most valuable of our invisible exports—our tourist trade. Natural scenery and historic monuments are both included in our assets for this purpose but above all it is the sum total of quiet beauty of Britain—what Man has made of the Land—which really counts.

*Stamp, L. D. (1937). Nationalism and Land Utilization in Britain. *Amer. Geog. Review*, 27, 1937.
See also Stamp, L. D. (1948). *The Land of Britain: Its Use and Misuse*, p. 438.
†Recently I have made an attempt (Stamp, 1954) to measure more precisely the relative outputs of different types of land by introducing a Potential Production Unit (P.P.U.) equivalent to the average output of ordinary good farm land of categories 2 and 4. Such land supports 0.4 stock unit per acre; it provides sustenance at our standard of living for one person per acre. Best lands have a ranking of 2 P.P.U.; poor lands such as No. 9 only 0.1 P.P.U. Thus a new town covering 5,000 acres built on the best land eliminates 10,000 P.P.U., on light sandy land (such as that underlying Bournemouth) only 500 P.P.U.

Possible Changes in Land Use in Great Britain

	1931-5					Possible future totals	
Use	Acreage	Per-centage	Possible changes in use	Revised per-ncetages	Use*	Per-centage	Acreage
(1) Intensive arable (market gardens)	1,100,000	2	No change	2	(1)	5	2,820,000
(2) Arable: farm crops	11,069,000	20	Intensive　3 No change　14.5 Orchards & fruit　1 Unproductive　1.5	(2)	27.5	15,500,000
(3) Orchards & fruit	261,000	0.5	No change	0.5	(3)	1.5	845,000
(4) Grassland (permanent)	18,967,000	33	Arable　10 No change　22 Houses with gardens　1	(4)	25	14,100,000
(5) Heathland Moorland and rough grazing	18,775,000	33	Arable　3 Grassland　3 No change　15 Forest　10 Houses with gardens　2	(5)	15	8,450,000
(6) Forest & Wood-land	3,219,000	6	No change	6	(6)	16	9,000,000
(7) Houses with gardens	1,720,000	3	No change	3	(7)	6	3,380,000
(8) Unproductive	1,399,000	2.5	No change	2.5	(8)	4	2,250,000
Total †	56,510,000	100		100		100	56,345,000

* The numbers in parenthesis indicate the land uses correspondingly numbered in the first column of the table.

† The slight discrepancy between the total acreage in 1931-5 and in the future is due to the rounding-off acreages as calculated from percentages in the last column.

BIBLIOGRAPHY

General

TANSLEY, A. G. (1939, 1949). *The British Islands and their Vegetation.* Cambridge: University Press.

TANSLEY, A. G. (1949). *Britain's Green Mantle.* London: Allen & Unwin.

STAMP, L. D. and BEAVER, S. H. (1933, 1937, 1941, 1954, 1963). *The British Isles: A Geographic and Economic Survey.* London: Longmans.

DARBY, H. C. (Editor) (1936). *An Historical Geography of England before A.D. 1800.* Cambridge: University Press.

STAMP, L. D. (1948, 1950, 1962). *The Land of Britain: Its Use and Misuse.* London: Longmans.

STAMP, L. D. (Editor) (1937-1947). *The Land of Britain: The Final Report of The Land Utilisation Survey of Britain.* In 92 parts or 9 volumes. London: Geographical Publications.

TROW-SMITH, ROBERT (1951). *English Husbandry from the Earliest Times to the Present Day.* London: Faber and Faber.

ERNLE, LORD (PROTHERO, R. E.) (1912, 1917, 1922, 1927, 1936). *English Farming Past and Present.* London: Longmans.

FREAM, W. (1892; 13th Edition 1949). *Elements of Agriculture.* London: Murray.

WATSON, J. A. SCOTT and MORE, J. A. (1924-1929). *Agriculture.* Edinburgh: Oliver and Boyd.

SEEBOHM, M. E. (1927, 1952). *The Evolution of the English Farm.* London: Allen and Unwin.

ORWIN, C. S. (1949). *A History of English Farming.* London: Nelson.

FRANKLIN, T. BEDFORD (1952). *A History of Scottish Farming.* London: Nelson.

VENN, J. A. (1923, 1933). *The Foundations of Agricultural Economics.* Cambridge: University Press.

EKWALL, W. (1936, 1940). *The Concise Oxford Dictionary of English Place Names.* Oxford: Clarendon Press.

STOKES, H. G. (1948). *English Place Names.* London: Batsford.

LEBON, J. H. G. (1952). *The Evolution of our Countryside.* London: Dobson.

Chapter 1—In the Beginning

GODWIN, H. (1948). Studies in the Post-Glacial History of British Vegetation. *Phil. Trans. Roy. Soc.*, 233, B.

GODWIN, H. (1951). Pollen Analysis (Palynology). *Endeavour*, 10 (a convenient summary).

GODWIN, H. (1956). *The History of the British Flora.* Cambridge: University Press.

ZEUNER, A. F. E. (1945). *The Pleistocene Period.* London: Ray Society.

ZEUNER, A. F. E. (1946). *Dating the Past: an Introduction to Geochronology.* London.

MATTHEWS, L. HARRISON (1952). *British Mammals.* London: Collins, New Naturalist.

MANLEY, GORDON (1952). *Climate and the British Scene.* London : Collins, New Naturalist.

PEARSALL, W. H. (1950). *Mountains and Moorlands.* London : Collins, New Naturalist.

TURRILL, W. B. (1948). *British Plant Life.* London : Collins, New Naturalist.

FLEURE, H. J. (1951). *A Natural History of Man in Britain :* London : Collins, New Naturalist.

SCOTT, RICHENDA (1948) and others. *Snowdonia.* London: Collins. New Naturalist.

FOX, CYRIL (1932, 1938, 1947). *The Personality of Britain.* Cardiff, Nat. Museum of Wales.

HAWKES, J. and C. F. C. (1949). *Prehistoric Britain.* London : Chatto and Windus.

CURWEN, E. C. (1938). *Air-Photography and the Evolution of the Cornfield.* London : Black (with valuable bibliography).

CRAWFORD O. G. S. (1924, 1928). Air Survey and Archaeology. Ordnance Survey Prof. Papers, No. 7.

CRAWFORD, O. G. S. and KEILLER, A. (1928). *Wessex from the Air.* Oxford : Clarendon Press.

CRAWFORD, O. G. S. (1953). *Archaeology in the Field.* London : Phoenix House.

CURWEN, E. C. (1937, 1954). *The Archaeology of Sussex.* London : Methuen.

Chapters 2 to 8

TREVELYAN, G. M. (1944, 1946, 1947). *English Social History.* London : Longmans.

TREVELYAN, G. M. (1926). *History of England.* London : Longmans.

THE PELICAN HISTORY OF ENGLAND, Vols. 1-8, London : Penguin Books

    RICHMOND, IAN (1953). *Roman Britain.*

    WHITELOCK, DOROTHY (1952). *The Beginnings of English Society.*

    STENTON, DORIS M. (1951). *English Society in the Early Middle Ages.*

    MYERS, A. R. (1952). *England in the Late Middle Ages.*

    BINDOFF, S. T. (1950). *Tudor England.*

    ASHLEY, MAURICE, (1952). *England in the Seventeenth Century.*

    PLUMB, J. H. (1950). *England in the Eighteenth Century.*

    THOMSON, DAVID, (1950). *England in the Nineteenth Century.*

BONHAM-CARTER, VICTOR, (1952). *The English Village.* London : Penguin Books.

GEORGE, DOROTHY, (1931). *England in Transition.* London : Penguin Books.

ORDNANCE SURVEY, (1931). Map of Roman Britain (2nd. Edition).

DARBY, H. C. (1952). *The Domesday Geography of Eastern England.* Cambridge: University Press. Also later volumes.

DARBY, H. C. (1940). *The Medieval Fenland and The Draining of the Fens.* Cambridge: University Press.

HOSKINS, W. G. (1950). *Essays in Leicestershire History.* Liverpool : University Press.

ROWSE, A. L. (1950). *The England of Elizabeth.* London : Macmillan.

TAYLOR, E. G. R. (1930). *Tudor Geography : 1485-1583.* London : Methuen.

TAYLOR, E. G. R. (1934). *Late Tudor and Early Stuart Geography : 1583-1650.* London : Methuen.

FUSSELL, G. E. (1947). *The Old English Farming Books 1523 to 1730.* London : Crosby Lockwood.

FUSSELL, G. E. (1950). *More Old English Farming Books 1731 to 1793.* London : Crosby Lockwood.

COLLINGWOOD, R. G. and MYRES, J. N. L. (1936). *Britain and the English Settlements*. Oxford: Clarendon.
HARRISON, W. *See page* 68
COLVIN, BRENDA (1948). *Land and Landscape*. London : Murray.
STROUD, DOROTHY, (1950). *Capability Brown*. London : Country Life.

Chapter 9—Some Facts, Figures and Fancies

STAMP, L. D. (1953). *Our Undeveloped World*. London : Faber and Faber. (American edition, 1952, entitled *Land for Tomorrow*. New York : Amer. Geogr. Soc.)
H.M.S.O. *Agricultural Statistics* 1950-1 *United Kingdom, Part I.*
  „ 1939-44 *England and Wales, Part I.*
  „ 1945-49 *England and Wales, Part I*
    and annually before 1939 and from 1950 onwards.
  Agricultural Statistics Scotland, annually before 1939 and from 1950 onwards.
    Combined volumes for war and post-war years.
UNITED NATIONS. *Demographic Yearbook, 1952* and later years.
UNITED NATIONS, FOOD AND AGRICULTURE ORGANIZATION. *Agricultural Statistics, 1952*.
STAMP, L. D. (1954). *The Under-Developed Lands of Britain*. London: Soil Association.

Chapter 10—Ceres

FRANKLIN, T. BEDFORD (1953). *British Grasslands from the Earliest Times to the Present Day*. London : Faber and Faber.
STAPLEDON, R. G. (1936). *A Survey of the Agricultural and Waste Lands of Wales*. London : Faber and Faber.
STAPLEDON, R. G. (1943). *The Way of the Land*. London : Faber and Faber.
MOORE, H. I. (1949). *The Science and Practice of Grassland Farming*. London : Nelson.
ARMSTRONG, S. F. (1917, 1921, 1937). *British Grasses and their Employment in Agriculture*. Cambridge : University Press.
SEMPLE, A. T. (FAO), (1952). *Improving the World's Grasslands*. London : Hill.
VAVILOV, N. I. (1949-50). *Origin of Cultivated Plants*. Chronica Botanica, 13.
HUNTER, M. (1951). *Crop Varieties*. London : Spon (Farmer and Stockbreeder).
PERCIVAL, JOHN, (1934). *Wheat in Great Britain*. London : Duckworth.

Chapter 11—Evolution in the Farmyard

DARWIN, CHARLES (1868). *The Variation of Animals and Plants under Domestication*, Vols. I and II. London : John Murray.
SAUER, CARL O. (1952). *Agricultural Origins and Dispersal*. New York : Amer. Geog. Soc.
ELLERMAN, J. R. and MORRISON-SCOTT, T. C. S. (1951). *Checklist of Palaearctic and Indian Mammals 1758 to 1946*. London : British Museum (Nat. Hist.).
HARTING, J. E. (1880). *British Animals Extinct within Historic Times*. London :
WALLACE, ROBERT (1885, 1889, 1893 etc.). *Farm Live Stock of Great Britain*. London : Crosby Lockwood.
GARNER, F. H. (1944). *The Cattle of Britain*. London : Longmans.
WHITEHEAD, G. KENNETH (1953). *The Ancient White Cattle of Britain and their Descendants*. London : Faber and Faber.
BOSTON, E. J. (Editor), (1954). *The Jersev Cow*. London : Faber and Faber.

THROWER, W. R. (1954). *The Dexter Cow*. London: Faber and Faber.
THOMAS, J. F. H. and others (1945). *Sheep*. London: Faber and Faber.
REID, J. W. (1949). *Pigs*. London: Spon.
GARNER, F. H. (1946). *British Dairying*. London: Longmans.
H.M.S.O. Cmd. 6494. *Report of the Committee on Hill Sheep Farming in Scotland (1944)*.

Chapter 12—Pomona

TAYLOR, H. V. (1936, 1945, 1946). *The Apples of England*. London: Crosby Lockwood.
TAYLOR, H. V. (1940). *Plums*. London: Crosby Lockwood.
GRUBB, N. H. (1948). *Cherries*. London: Crosby Lockwood.
STAMP, L. D. (1948, 1950, 1962). *The Land of Britain: Its Use and Misuse*, Chapter VI.
LANGLEY, BATTY (1729). *Pomona, or The Fruit Garden*. Illustrated.
BEALE, J. (1724). *Herefordshire Orchards*.
BUSH, RAYMOND (1943). *Tree Fruit Growing I and II*. London: Penguin Books.
BUSH, RAYMOND (1945). *Frost and the Fruit Grower*. London: Cassell.

Chapter 13—Sylva

TANSLEY, A. G. (1939), 1949. *British Islands and their Vegetation*. Cambridge University Press.
BRIMBLE, L. J. F. (1948). *Trees in Britain, Wild, Ornamental and Economic*. London: Macmillan.
TAYLOR, W. F. (1945). *Forests and Forestry in Great Britain*. London: Crosby Lockwood.
EDLIN, H. L. (1944.) *British Woodland Trees*. London: Batsford.
EDLIN, H. L. (1956). *Trees, Woods and Man*. London: Collins, New Naturalist.
UNITED STATES DEPARTMENT OF AGRICULTURE, (1949). *Trees: Yearbook of Agriculture, 1949* (for details of American Species).
CANADA DEPARTMENT OF MINES AND RESOURCES, (1917-1949). Native Trees of Canada. *Bull No. 61*, 4th Ed. 1949 (for details of Canadian trees).

Chapter 14—Trees of the Forest, 1919

FORESTRY COMMISSION. *Census of Woodlands 1947-1949*. Census Report No. 1. H.M.S.O. 1952.
Annual Reports of the Forestry Commission
TAYLOR, W. L. (1945). *Forests and Forestry in Britain*. London: Crosby Lockwood.
H.M.S.O. Cmd. 6447 (1943). *Post War Forest Policy*.
FORESTRY COMMISSION. *Britain's Forests:*—Rheola (1949), Coed y Brenin (1950). Forest of Ae (1948), Culbin (1949), Kielder (1950), Tintern (1950), Glentress (1953), Strathyre (1951), Thetford Chase (1951).
FORESTRY COMMISSION: NATIONAL FOREST PARK GUIDES:—Forest of Dean (1947), Snowdonia (1948), Glen More (1949), Glen Trool (1950), Hard Knott (1949), Argyll (1947).
FORESTRY COMMISSION GUIDE New Forest (1951, revised 1952).
FORESTRY COMMISSION Guide to the National Pinetum and Forest Plots at Bedgebury (1951).
FORESTRY COMMISSION The Dedication of Woodlands.

Chapter 15—Guests

LOUSLEY, J. E. (Ed.) (1953). *The Changing Flora of Britain.* Arbroath: Buncle.

SALISBURY, Sir E. J. (1961). *Weeds and Aliens.* London: Collins, New Naturalist.

SHORTEN, M. (1948). *New. Nat. J.*1.: 42-46.

MIDDLETON, A. D. (1931). *The Grey Squirrel.* London: Sidgwick and Jackson.

MOFFAT, C. B. (1938). The Mammals of Ireland. *Proc. Roy. Irish Acad.* 44: 61-128 (p. 103).

DARWIN, C. (1868) op. cit.

TICEHURST, N. F. quoted in Witherby's *Handbook of British Birds.* vol. 3: 177.

MATTHEWS, L. HARRISON (1952). *British Mammals.* London: Collins, New Naturalist.

SMITH, M. (1951). *The British Amphibians and Reptiles.* London: Collins, New Naturalist.

VEALE, ELSPETH M. (1957). The Rabbit in England. *Agric. Hist. Rev.,* 5: 85-90.

Chapter 16—The Nineteenth Century

ASTOR, VISCOUNT and MURRAY, K. A. H. (1933). *The Planning of Agriculture.* London: Oxford University Press.

WHETHAM, E. H. (1952). *British Farming 1939-49.* London: Nelson.

CLARK, F. LE GROS and PIRIE, N. W. (Editors) (1951). *Four Thousand Million Mouths.* London: Oxford University Press.

STAMP, L. D. (1960). *Our Developing World.* London: Faber and Faber.

WIBBERLEY, G. P. (1950). Changes in Agriculture and their Effect on the Countryside. *Town and Country Planning School, 1950.*

HENDERSON, H. C. K. (1952). Agriculture in England and Wales in 1801. *Geog. Jour.* 118: 338-345.

Chapter 17—Towards National Planning, 1895-1947

WATSON, J. A. SCOTT, (1938). *The Farming Year,* London: Longmans.

HALL, A. DANIEL, (1911). *Reconstruction and the Land.* London: Macmillan.

H.M.S.O. Cmd. 6378. *Report of the Committee on Land Utilisation in Rural Areas* (1942).

DOWER, JOHN, (1945). *National Parks in England and Wales.* London: H.M.S.O. Cmd. 6628.

H.M.S.O. Cmd. 7121. *Report of the National Parks Committee* (1947).

H.M.S.O. Cmd. 7122. *Conservation of Nature in England and Wales* (1947).

A series of maps on the scale of 1 : 625,000 (approximately 10 miles to one inch) is published by the Ordnance Survey as the National Planning Series, each in two sheets (Northern and Southern) to cover England, Wales and Scotland. Important ones include Relief, Geology, Rainfall, Land Utilisation, Types of Farming, Vegetation (England and Wales only), Land Classification, Coalfields, Administrative Areas, Population and Population Changes.

Chapter 18—The Second Elizabethan Age and the Future

ELLISON, W. (1953). *Marginal Land in Britain.* London: Geoffrey Bles.

STAMP, L. D. (1962). *The Land of Britain: Its Use and Misuse,* 3rd Edition, Chapter XXII. London: Longmans.

HOSKINS, W. G. and STAMP, L. D. (1963). *The Common Lands of England and Wales.* London: Collins, New Naturalist.

NATURE CONSERVANCY, (1960). *The First Ten Years.*

BEST, R. H. and COPPOCK, J. T. (1962). *The Changing Use of Land in Britain.* London: Faber.

INDEX

Note: All authors quoted in the text are indexed but not the titles of works.
Incidental references to places or regions have not been indexed.